D0875932

Envisioning the City

The Kenneth Nebenzahl, Jr., Lectures in the History of Cartography

Published for the Hermon Dunlap Smith Center for the History of Cartography The Newberry Library

Series Editor, David Woodward

Maps: A Historical Survey of Their Study and Collecting
by R. A. Skelton
(1972)

British Maps of Colonial America
by William P. Cumming
(1974)

Five Centuries of Map Printing
edited by David Woodward
(1975)

Mapping the American Revolutionary War
by J. B. Harley, Barbara Bartz Petchenik, and Lawrence W. Towner
(1978)

Series Editor, David Buisseret

Art and Cartography
edited by David Woodward
(1987)

Monarchs, Ministers, and Maps
edited by David Buisseret
(1992)

Rural Images
edited by David Buisseret
(1996)

Series Editor, James Akerman

Envisioning the City
edited by David Buisseret
(1998)

Envisioning the City

Six Studies in Urban Cartography

Edited by David Buisseret

The University of Chicago Press ❧ Chicago & London

DAVID BUISSERET is the Garrett Professor of History,
University of Texas at Arlington.

THE UNIVERSITY OF CHICAGO PRESS, CHICAGO 60637
THE UNIVERSITY OF CHICAGO PRESS, LTD., LONDON

Printed in the United States of America

07 06 05 04 03 02 01 00 99 98 1 2 3 4 5

ISBN: 0-226-07993-7 (cloth)

Library of Congress Cataloging-in-Publication Data

Envisioning the city : six studies in urban cartography / edited by David Buisseret.
p. cm. — (The Kenneth Nebenzahl, Jr., lectures in the history of cartography)
Includes bibliographical references and index.
ISBN 0-226-07993-7 (cloth : alk. paper)
1. Cities and towns—Maps. I. Buisseret, David. II. Series.
G140.E58 1998
912'.19732—dc21 97-37158
CIP

Contents

Editor's Note

THE ESSAYS published here are revised versions of the lectures delivered in the tenth series of Kenneth Nebenzahl, Jr., Lectures in the History of Cartography, held at the Newberry Library, Chicago, in November 1991. This series was entitled "Profiling the City," but having considered the observations of the very helpful readers provided by the University of Chicago Press, as well as the central theme of most of the contributions, we decided to call the book "Envisioning the City." The theme is vast, and there are no doubt important aspects of urban cartography that we have neglected. But we hope to have brought out the essentially subjective element present in this kind of mapping, as in all others, and we shall be delighted if our efforts lead to fuller and deeper studies of these interesting and generally neglected problems.

This is now the tenth series of Nebenzahl Lectures, and the eighth to appear in printed form. We should like to thank the many institutions which allowed us to reproduce material from their collections, and the many librarians who helped us in our research. Thanks are due as well to Maria Veda, my graduate assistant, who largely prepared the index. Finally, we also have to thank Mr. and Mrs. Nebenzahl, who founded these lectures and continue both to support them and to make useful suggestions concerning future themes.

Introduction

David Buisseret

Town plans as a genre are curiously neglected in the literature. They are probably the oldest form of map and have been produced in great numbers, yet very few studies have been devoted specially to them. This present collection of essays does not, of course, set out fully to survey the field, but it does aim to develop some themes that could eventually be included in a more complete work; almost all these themes concern in one way or another the theory and practice of representation, or the manner in which urban areas have been envisioned.

The fullest and most recent consideration of town plans is the short book produced by James Elliot in association with a British Library exhibition of 1987, called *The City in Maps: Urban Mapping to 1900* (London, 1987). This work, betraying its origins, is rather anglophone in its choice of material, but it does give an overview of urban mapping and has useful bibliographies. Several general histories of mapping have chapters on town plans: P. D. A. Harvey considers them in *The History of Topographical Maps* (London, 1980), as does A. G. Hodgkiss in *Understanding Maps* (London, 1981). More recently, Jan Mokre offers us "Grundriss contra Aufriss: Die Stadt in der Kartographie" in *Kartographische Zimelien* (ed. Franz Wawrik et al. [Vienna, 1995]. There are also several useful chapters in the magisterial volumes of the still emerging Harley-Woodward *History of Cartography* (Chicago, 1987–).

Curiously, two of the most perceptive and wide-ranging contributions to our theme appeared as articles written twenty years ago. John Pinto's "Origins and Development of the Ichnographic City Plan" brilliantly evokes the emergence of these plans in Renaissance Europe, and Jürgen Schulz's work, "Jacopo de' Barbari's View of Florence," offers a masterly survey of the way in which this plan fits into the general development of European cartography up to this time. The tradition of making city plans goes back much further in Asia than in Europe, and one very suggestive recent article is Michel Cartier's "Les métamorphoses imaginaires de la Venise chinoise"; there must be many other such specialized articles in the regional literature.[1]

The Characteristics of City Plans

Since most scholars now agree that the content of all maps is subjective, we might take the view that any area of the earth's surface contains as many features of interest as any other; it would thus be possible, for instance, to map an area of desert with minute attention to each layer of stone and each sage bush. Still, most people would agree that in practice towns contain an altogether exceptional density of features that can be mapped: streets, buildings, walls, bridges, and so forth. These features are so concentrated that city plans have had to develop special ways of showing them; such maps are also often multilayered in a vertical dimension, and this introduces a further representational complication.

Vertical Plans

The oldest way of delineating towns is through use of a more or less vertically seen plan; maps of this kind go back to the earliest times in the Middle East, where they were drawn not only on forms of paper but also on tablets. As Cornelis Koeman puts it, with characteristic force and brevity, "alle alten Zivilisationen haben Ortspläne gekannt" [all old civilizations have known location maps].[2] In chapter 1, Nancy Shatzman Steinhardt reflects on the significance of such plans in Chinese culture, where they seem to have played an unusually important role.

In the classical West, one of the most famous of these plans was the huge "Forma Urbis Romae," a great plan of the city of Rome incised there on a wall; this seems to have been only the most famous of a number of such plans, which emperors liked to commission in celebration of their rule. The ancient technique of drawing city plans was probably not lost after the fall of the Roman empire; for instance, Charlemagne is known to have had at his court plans of both Rome and Constantinople.[3] It certainly came to renewed life in medieval Italy, where from the tenth century onward plans are known of cities such as Verona, Rome, Venice, and Padua.[4] This primarily Italian skill had spread by 1421 to Vienna in the shape of a famous plan that had numerous counterparts during the sixteenth century.

At that time, too, measured plans began to be drawn for new purposes. Soldiers produced extremely accurate plans for towns to be protected by the new bastioned traces (see the chapter by Martha Pollak), and many towns were included in the estate maps that were drawn for improving landlords, particularly in England. For the first time, huge surveys of all the towns in a region were commissioned, like the one initiated for Philip II of Spain by Jacob van Deventer in the 1570s.[5] Many of these plans eventually found their way into the first great printed treasury of city images, the *Civitates Orbis Terrarum* published by Georg Braun and Franz Hogenberg between 1572 and 1617.[6]

Further printed collections of city plans appeared during the seventeenth and eighteenth centuries, and towns were also included in the great national surveys of the nineteenth century. At that time, too, specialized thematic city plans began to appear; the most widespread of them were probably the fire insurance plans, which showed urban areas with hitherto unimagined detail with respect to individual buildings. In the twentieth century, electronic plans have supplanted all these older types; they offer the great advantage of being able to show instantly a variety of features—sewers, communications networks, educational institutions, and so forth—on the same base map, sometimes in combination with each other. This allows city planners to call up selections of spatial information that would formerly have been very difficult to generate.

Bird's-Eye Views

Although the vertical plan is the oldest and probably the most common way to show a city, it has in the Western tradition long been accompanied by the bird's-eye view, also known as the oblique or cavalier view. One of the earliest such views was the magnificent image of Florence produced in 1475. It demonstrates the advantages of the type: a view of this kind not only gives some idea of the town's layout (better shown on a vertical plan), but also allows the artist to convey some impression of the vertical dimension, by indicating the tallest (and generally most significant) buildings.

Many cities came to be shown in this way during the sixteenth century. Probably the finest view was the one provided of Venice by Jacopo de' Barbari in 1500; it was followed by similar very large views of Augsburg (1521), London (1553), and Bruges (1562), to name only the most famous. Many such views also appeared in Braun and Hogenberg's *Civitates Orbis Terrarum,* and in the seventeenth-century collections by authors such as Matthaeus Merian and Wenceslaus Hollar. Madrid was brilliantly shown by Pedro Teixeira in 1656, and the genre continued into the eighteenth century, with works such as the Turgot plan of Paris in 1734 and Joseph Hubner's views of Prague and Vienna.[7]

During the nineteenth century the bird's-eye view crossed the Atlantic with great success, becoming the prime means by which the new cities of North America set off their appeal.[8] During the twentieth century, such views have become somewhat less common, but they are still used when a cartographer needs simultaneously to give an idea of the general layout of a city and also to identify its principal monuments. The most remarkable practitioner of these views is no doubt Hermann Bollmann, whose minutely detailed work will allow future historians to visualize the exact shape of cities such as Hamburg and New York. The chapters here by Gerald Danzer and Richard Kagan demonstrate the use of the bird's-eye view in arguing the case for new developments and for civic excellence.

Profiles and Prospects

Towns may be viewed not only directly from above, as in a plan, or obliquely from above, as in a bird's-eye view, but also from ground level. "Profiles" of this kind have their counterparts in other areas of cartography; "landfalls," for instance, showing the land horizon as seen from the sea, were often used in early modern nautical charts, and mountain ranges are often shown in profile, with the peaks and passes easily identified. Sometimes, too, cities with skyscrapers are still shown in this way, for all these subjects share the common characteristic of having exceptionally tall features.

When the first printed books with city plans began appearing in Europe in the late fifteenth century, many of the towns were shown in profile, with their huge churches prominent—perhaps overprominent—above their city walls; many such views may be found in Hartmann Schedel's *Liber Chronicarum* of 1493. This type of delineation continued to be used into the sixteenth century, with splendid examples such as Hans Lautensack's 1552 elevation of Nuremberg, or the many profile views in Braun and Hogenberg's *Civitates Orbis Terrarum.* A view of this kind gives a particularly strong impression of a city with high monuments to a traveler who expects to arrive by foot, with the city panorama slowly unfolding as the approach is made; there are some fine examples from eighteenth-century England, where such views were often known as "A Prospect of. . . ."[9]

Profiles gradually lost ground during the sixteenth century to the plan and bird's-eye view, for the former could give a better image of the general layout and the latter was almost as useful for distinguishing the main buildings on the skyline. Profiles are rarely drawn today, but they often appear as photographs when authors want to convey an impression—say, of arriving in New York harbor, with its magnificent skyline.

Modeling the City

So far we have considered ways of delineating the city on paper or on stone. But if, as we believe, a map is a representation of a locality, and not necessarily only a graphic representation, then we ought as well to consider models. They do not exist for all times and places, though many may yet be found for cultures in which we presently know nothing of them. But there is a long and vigorous tradition of using models to portray towns in early modern Europe, a tradition that seems to derive from the medieval practice of modeling large buildings before they were constructed.

The royal French collection ("Les Plans-Reliefs"), now housed mostly at the Musée des Invalides in Paris, is the chief assembly of maps of this kind, but others exist in other European countries because, during the eighteenth century, it became almost mandatory for forward-looking monarchs to commission a collection of such models. As we gaze at the Paris collection, we are impressed both by the immense labor that went into building them and by their inherent fragility. But we cannot as well help reflecting that they are an ideal way to seize the main outlines and then the detail of a city, combining the virtues of plan, bird's-eye view, and profile.

In our century, many cities have seen the advantage of providing their visitors with a model, sometimes old and sometimes newly made, sometimes showing the city as it is and sometimes as it once was. There is no better way to gain an instant acquaintance with the main outlines of an unfamiliar town, though the initial cost is high. This immediacy of information has long been appreciated by military commanders, for whom cost is not generally of prime importance. Louis XIV prized his models, but so did Napoleon Bonaparte, who, as we shall see, thought them "the best kind of map." During the Second World War, the Allied commanders made extensive use of town models in order to familiarize both land and air forces with their targets; so did the Japanese, who made a detailed model of Pearl Harbor before attacking it. One way and another, the model, for all its material problems, is absolutely the best way to convey the most spatial information about a city to an uninformed viewer in the shortest possible time.

Representing the City

In the chapters that follow, the authors describe a variety of ways of envisioning urban areas, though all their examples fall into one of the four categories that we have described: vertical plan, bird's-eye view, profile, or model. The interest of the analysis lies in the way that all the authors of the various types of representation try, as Gerald Danzer puts it, to show the city "the way people want to see it."

Thus Nancy Shatzman Steinhardt shows that the mapping of Chinese cities was "an art among elite arts," concerned to perpetuate a certain ideal of the city, without close regard to what it was or would be actually on the ground. Naomi Miller, too, shows that while many of her Ptolemaic maps may easily be related to current urban layout, they were made with specific representational aims in view, now military, for instance, and now didactic. Richard Kagan brings the contrast between what actually exists and "the way people want to see it" into the center of his argument, distinguishing between images of the city as *urbs,* an architectural entity, and what he calls "metonymic" representations of the *civitas,* designed perhaps to "ennoble" the city, to draw attention to its defenses, to emphasize its Christian aspect, or indeed to exalt the authority of the king. Even the essays by Martha Pollak and David Buisseret, on the apparently objective military representation of cities, bring out subjective elements of the work, whether in fortification plans drawn partly to inspire terror or in models assembled in great numbers to compel admiration. Finally, Gerald Danzer addresses the question of how Daniel Burnham engaged a distinct representational style in order to "sell" his version of the new Chicago. Burnham had recourse not only to very elegantly conceived vertical plans but also to delicate, atmospheric watercolor bird's-eye views, all produced so that the reader would be not only convinced by the text but seduced by the image.

In the course of these chapters the authors raise various subsidiary issues of great interest: How, for instance, does the planning of the palace within the Chinese walled city compare with the placing of the citadel within the Western bastioned trace? Or, again, what are the stages by which the literate European public became aware of the world's great cities, as shown not only in the frequently quoted work of Braun and Hogenberg but also in the voluminous fortification works containing city plans? Yet despite the variety of issues these essays raise, they all agree in demonstrating one central point: No matter what technical means of representation is adopted, the actual envisioning of the city owes everything to the motives of its patrons, as reflected in the technique of the plan makers.

Notes

1. See John Pinto, "Origins and Development of the Iconographic City Plan," *Journal of the Society of Architectural Historians* 35 (1976): 35–50; Jürgen Schulz, "Jacopo de' Barbari's View of Venice," *Art Bulletin* 60 (1978): 427–47; and Michel Cartier, "Les métamorphoses imaginaires de la Venise chinoise; la ville de Suzhou et ses plans," *Revue de la Bibliothèque Nationale* 48 (1993): 2–9.

2. Cornelis Koeman, "Die Darstellungmethoden von Bauten auf alten Karten," in *Land- und Seekarten im Mittelalter und in der frühen Neuzeit,* ed. Cornelis Koeman (Munich, 1980), 145.

3. A. J. Grant, ed., *Einhard's Life of Charlemagne* (London, 1907), 54.

4. Jan Mokre, "Grundriss contra Aufriss: Die Stadt in der Kartographie," in *Kartographische Zimelien,* ed. Franz Wawrik et al. (Vienna, 1995), 19–27.

5. See Robert W. Karrow, ed., *Mapmakers of the Sixteenth Century and Their Maps* (Chicago, 1993), 142–58.

6. Most accessible in the facsimile edited by R. A. Skelton (Amsterdam, 1980).

7. For the latter, see Mokre.

8. See the many publications of John W. Reps, including particularly *Views and Viewmakers of Urban America* (Columbia, 1984).

9. For instance, Robert Morden, *A Prospect of London* (London, c. 1700).

ONE

Mapping the Chinese City: The Image and the Reality

Nancy Shatzman Steinhardt

Everyone has seen a map of a United States city with dotted lines printed in anticipation of roads still under construction. Some have had the experience of using such a map at one time—following a thoroughfare to the point where the dotted lines (and construction) begin—and returning a few years later to find the road complete and open. The same practice has been employed in maps of the rapidly modernizing Chinese capital, Beijing, for which maps showing its third and fourth ring roads could be found several years before they were complete and open for traffic. Maps of Taibei in the 1960s, when the Taiwanese capital anticipated its role as a showcase of East Asian modernization, showed straight broad roads in neighborhoods that in actuality consisted of narrow lanes and alleys. Unlike the United States or Beijing city maps, however, the Taibei maps did not indicate new construction underway. Rather, the image on the maps was the projected vision of city planners.[1]

The 1960s maps of Taibei borrowed from historical precedent in the publication of a city plan so different from the actual urban configuration. The manufacture of a city plan without regard to the form that would emerge from the physical evidence was an age-old Chinese practice whose purpose, moreover, was not related to future expectations. Rather, lines that mark streets or walls that never existed and that never need exist represent a determination to envision a contemporary city, whatever its actual configuration, in idealized and presumed archaic form. The phenomenon is especially suprising when one considers the length of the history of cartography in China and the very early capabilities of rendering the physical likeness of a city or larger region in lines drawn according to scale.

The best proof of how different the mapped image of Chinese cities can be from the actual cities is available for urban centers whose plans can be confirmed by excavation. Most of these excavated cities were at one time capitals. Since the majority of this archaeological activity has occurred in the last few decades and is published primarily or exclusively in Chinese, as is most of the literature on Chinese planning, a brief summary of the history of Chinese maps of capitals follows.

Royal Chinese City Planning

Chinese discussions of the plans of rulers' cities, including many of the most recent ones, begin with the focus on a key passage in a classical text that offers a prescription for an ideal state capital.[2] Actually it is *the* ideal state capital, for just as one familiar with Chinese elite culture is not surprised by the homage paid to a text, neither does one give a second thought to the fact that once something appears in *the* literature, it can quickly become the irrefutable text. The discussion of what is called Wangcheng, literally, "ruler's city," is found in a late, first millennium B.C. section of *The Rituals of Zhou,* whose relevant lines for city planning may survive from a slightly earlier version. It is generally agreed that the ritual and related matter are those of the Zhou people, who ruled China from the end of the second millennium through the middle of the third century B.C. In other words, chronologically the text has the sort of pedigree sought by later promoters of Chinese ancient culture at the emperor's court. The famous passage reads:

> The *jiangren*[3] builds the state, leveling the ground with the water by using a plumb-line. He lays out posts, taking the plumb-line (to ensure the posts' verticality), and using their shadows as the determinators of a mid-point. He examines the shadows of the rising and setting sun and makes a circle which includes the mid-points of the two shadows.
>
> The *jiangren* constructs state capitals. He makes a square nine *li* [1 *li* = about ½ km] on each side; each side has three gates. Within the capital are nine north-south and nine east-west streets. The north-south streets are nine carriage tracks in width. On the left (as one faces south, or, to the east) is the Ancestral Temple, and to the right (west) are the Altars of Soil and Grain. In the front is the Hall of Audience and behind the markets.[4]

The earliest known illustration of the Wangcheng described in this passage is from a fifteenth-century encyclopedia.[5] The other famous illustration of the ruler's city was published in 1676 (figure 1.1). From the words and pictures we can isolate seven principles of Chinese imperial planning: (1) Preparation of the site requires a reconciliation between the man-made forms and nature (often accomplished under the direction of a geomancer). (2) Construction begins with the determination of a midpoint.

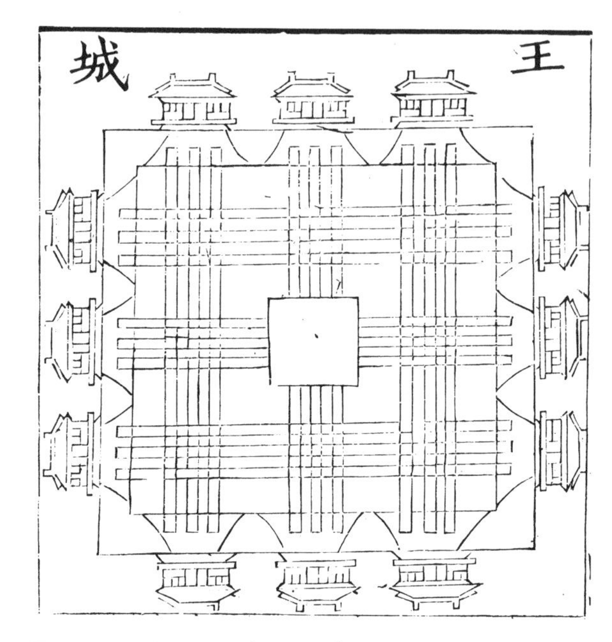

FIGURE 1.1. *Wangcheng* (Ruler's City). Nalan Chengde, ed., *Sanli tu* [Illustrated *Three Ritual Classics,* including *Rituals of Zhou*], 1676, pt. 1, *juan* 4/2b.

(3) Construction proceeds from this midpoint to the outer wall boundaries (ideally completed, but in reality at least initiated, before interior building commences). (4) The boundary of a city is a wall and that wall's shape is a square. (5) Three gates should provide access to and from the city through each outer wall. (6) Major arteries of the city cross at right angles. (7) Every imperial city should contain a temple to the ruler's ancestors on the east, altars for sacrifices to soil and grain on the west, a hall of audience at the south and facing south, and markets in the north. Several of these features are shown in an illustration from an early twentieth-century edition of the *Shu jing* (Book of history) (figure 1.2). The area of the ruler's city that contains his residential palaces is enclosed and has various Chinese names. It is best referred to in English as the

FIGURE 1.2. *Divining the Capital at the Jian and Chan Rivers.* From Sun Jianai et al., *Shu jing tu shuo* [Illustrations and notes to the *Shu jing* (Book of History)], 1906, *juan* 33/6a.

"palace city" (*gong cheng*). The space between the palace city and the outer wall can be called the "outer city."[6]

In the 1950s the vicinity of modern Luoyang in Henan province that included the site of Wangcheng was excavated.[7] According to texts, the Zhou ruler's city should have been built in about 1100 B.C. Inscriptions on bronze vessels tell us that two walled cities, Wangcheng and

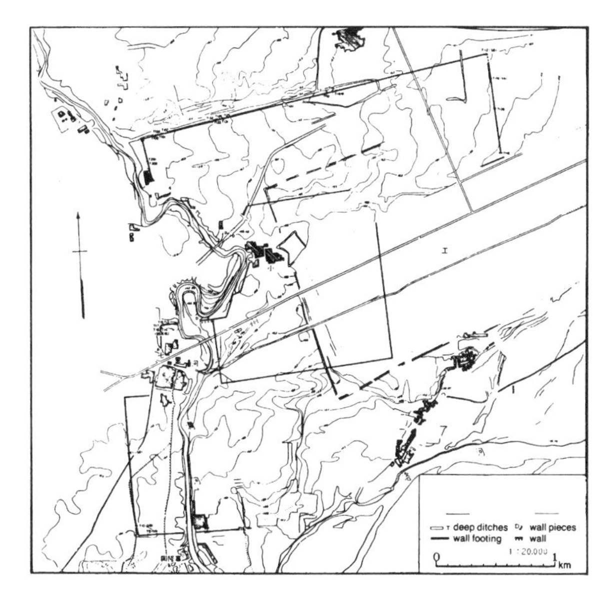

FIGURE 1.3. Remains at Luoyang from the Zhou and Han dynasties. After *Kaogu xuebao* [Archaeology Journal] no. 3 (1959), between pp. 16 and 17.

Chengzhou, were located in the general vicinity of modern Luoyang at this time. Excavation uncovered wall sections from both the Zhou and the Han (206 B.C. to A.D. 220) dynasties (figure 1.3), but none bore explicit resemblance to the capital described in the *Rituals of Zhou* or to the fifteenth- or eighteenth-century plans of it.

At least two first millennium B.C. Chinese cities did have plans with palace cities that can be described as in the center. One of them was the capital of the state of Lü where Confucius was a statesman at the beginning of the sixth century B.C., and another was in Shanxi province. However, excavation has shown that plans of capitals of China's scores of states during the first millennium B.C. also came in other varieties. In fact, three city plans dominated Chinese urbanism in the territory of Zhou rule during the first millennium B.C. In addition to cities with palace complexes near the center, the second city plan had a walled enclosure at approximately the north center of a larger wall, and the third had adjacent cities that bordered each other along at least one plane. When one of the two walled areas was smaller, it could function as a palace city. Other times, a concentration of buildings in one sector of one enclosure was the palace area. Excavation of multiple wall pieces suggests that Luoyang of the early Zhou was an example of the third and most prevalent type of first millennium B.C. city. Even though only the two of the scores of excavated Zhou cities are examples of a *Rituals of Zhou*–style plan, scholars continue to write that the classical text is the basis of all Chinese city planning.

Capital city planning of the first major dynasty that followed Zhou, the Han, shows that Chinese rulers were becoming convinced of the merit of one palace city as the focus of imperial activity. Whereas the most influential ruler of the short-lived Qin dynasty (255–206 B.C.) had at least seven palace complexes, and rulers of the first part of the Han (206 B.C.–A.D. 8) had five, the second Han capital had two palace areas, only one of which was used by the ruler at a time. In the third through the sixth centuries, an age of disunion in China, one can again find each of the three plans suggested by excavation of Zhou cities. Capitals of six dynasties centered around the modern city Nanjing in southeastern China resembled the *Kaogong ji* prescription for Wangcheng. Double cities were built as far north as what is today the Inner Mongolian Autonomous Region and in Sichuan province of southeastern China. The city plan that emerges as extremely important during these three centuries has its palace area in the north center. It was constructed at both Ye and Luoyang in Henan province. More important, the scheme was the forerunner of the greatest planned city in medieval East Asia, the capital of Sui-Tang China at Daxing-Chang'an (modern Xi'an) from the end of the sixth through the beginning of the ninth century (figure 1.4). This city plan spread with Tang Chinese culture to Japan, where no fewer than five cities were modeled after it, and to northeast continental Asia under Korean and Bohai rule; it is even the source of the plan of a thirteenth–fourteenth-century Mongolian prince's city at Yingchanglu in the Inner Mongolian Autonomous Region.

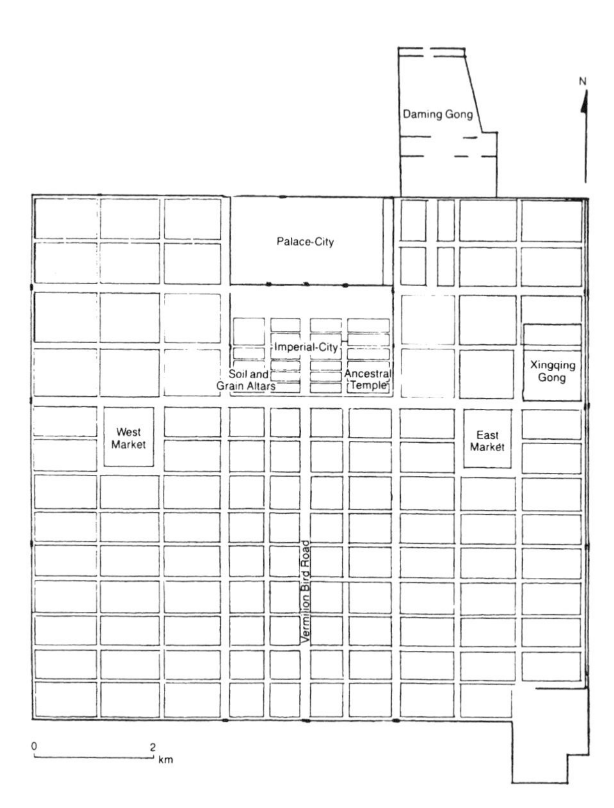

FIGURE 1.4. Plan of Tang Chang'an. After *Wenwu* [Cultural Relics] no. 9 (1977), p. 2.

The fall of Tang marked the beginning of three centuries of rule by non-Chinese dynasties. The process began along China's northern fringe, but culminated in the thirteenth century with the

Mongolian occupation of the entire country of China, itself only a small part of the Asian empire. Many of the more than ten capital cities of non-Chinese northern rulers were examples of double cities, but with a particular function. Here textual evidence is insightful, for it suggests the walls that divided northern from southern cities were also a means of segregating the non-Chinese ruling population from the rest of the ethnic groups living in the city. Moreover, the rulers lived in a more fortified part of the city, farther from the most likely direction of attack.

The conquest of China by the Mongols and the establishment of the Mongolian empire in China by Khubilai Khan in the 1260s marked more than the culmination of the accession of non-Chinese dynasties to the Chinese throne. The Mongolian capital in East Asia was the first city for which it can be documented that the design was consciously selected to follow the *Kaogong ji* passage; in fact, it followed the text more closely than any imperial city built in China before or after it (figure 1.5).[8] This was in spite of the fact that Khubilai had been born a steppe nomad.

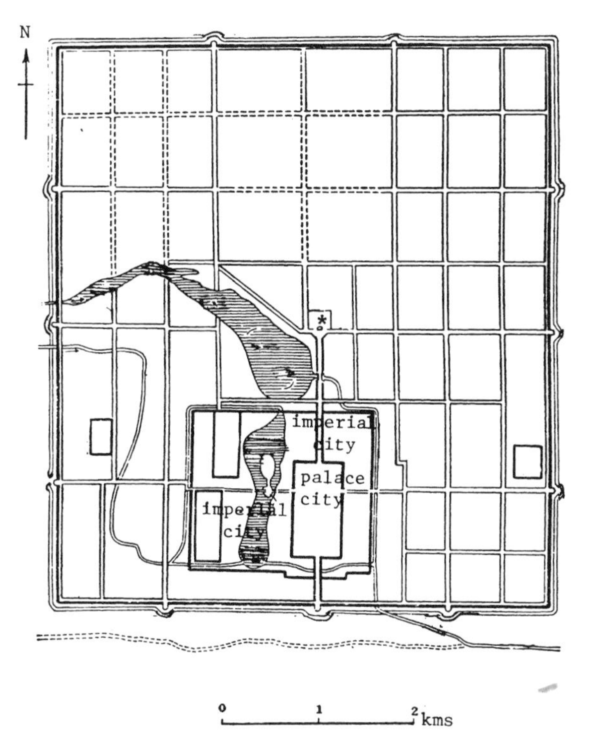

FIGURE 1.5. Plan of Yuan Dadu. After *Wenwu* no. 9 (1977), p. 5. The center marker is represented by a star.

The most significant and still unique element in the plan of Khubilai's capital Dadu was the placement of a "center marker" (*zhongxin zhi tai*). This engraved stone, shown in figure 1.5 as a small square at the southeastern terminus of a diagonal street, would have been midway between the north and south and the east and west outer walls. It was put in place before wall construction began. Thus building order was a second unique association between Dadu and Wangcheng. (Building had been in reverse order at Sui-Tang Chang'an.) Third, the ancestral temple, one of the four buildings or building complexes specified in the *Kaogong ji* passage as imperative to the Chinese capital, was begun in 1264 (orders to construct it were given in 1263), three years before construction of the outer city wall.[9]

The importance of the center marker being what it was, one may be surprised that the palace city of Dadu was not built around it. The reason was that in China, when it came to actual construction, pragmatism could supersede the written word. At Khubilai's city the flow of water into, around, and out of the capital must have been deemed more essential to the city's life than its textual justification.

Ironically, it was the plan of the Mongolian capital that was responsible for the organization of space in later Beijing, constructed by the native Ming dynasty at the beginning of the fifteenth century on the ruins of Dadu. The

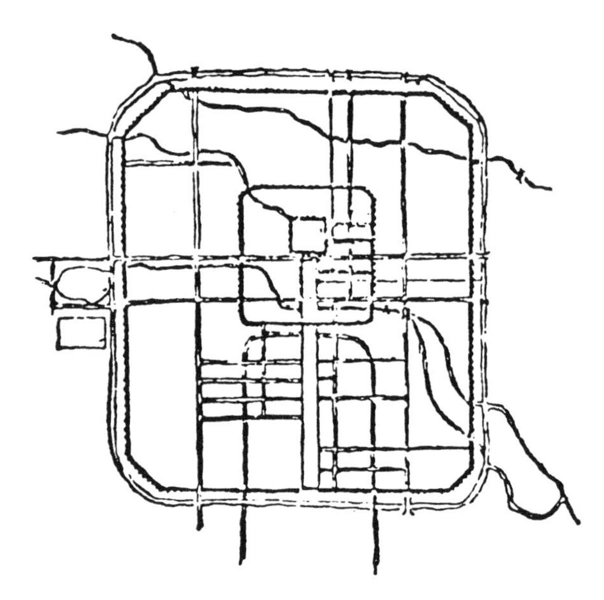

FIGURE 1.6. Plan of Northern Song Bianliang. After *Wenwu* no. 9 (1977), p. 3.

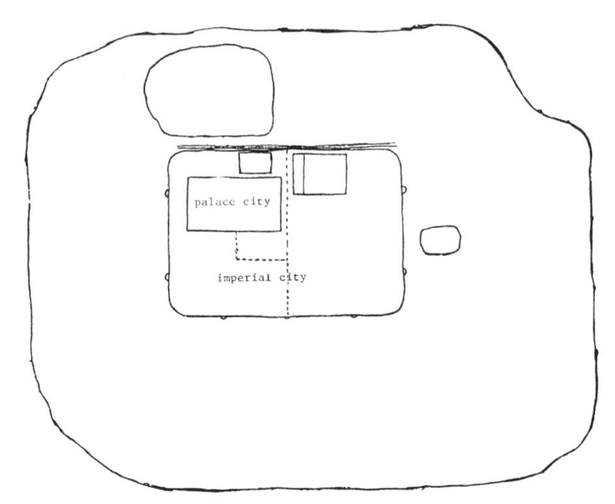

FIGURE 1.7. Plan of Northern Song Bianliang. After Yao Jiazao, *Zhongguo jianzhu shi* [History of Chinese architecture] (Shanghai, 1933), fig. 96.

Ming scheme gave way to seventeenth- and eighteenth-century Beijing and ultimately to the plan of today's capital. For no Chinese city before or after Dadu, however, can the kind of documentation that links Dadu and the *Rituals of Zhou* text be found.

EVIDENCE OF THE IMAGE AND THE REALITY IN MAPS OF SONG CAPITALS

MORE VARIETY IS FOUND in maps of the tenth- to twelfth-century Northern Song capital at Bianliang (modern Kaifeng) in Henan province than in maps of any other Chinese capital. Figure 1.6 is a plan that shows three concentric walled areas (palace, imperial, and outer cities), the wide approach from the south to them, and markedly cut-off corners. Texts confirm that the Northern Song capital did have palace, imperial, and outer cities. However, the written records explicitly state that the palace area was located northwest with respect to the outer wall.[10] In addition, unlike Dadu or other capitals to which populations of conquered peoples were moved or resettled and for which a founding population figure could be anticipated, the open society (by Chinese standards) and free trade that Song China offered, motivated people to flood into Bianliang. Chinese people have always desired to live inside of walls, so, in response to rapid growth, the Northern Song capital was walled, walled again, and walled yet again.[11] Not only did building order sharply contrast with the stipulations of the *Rituals of Zhou,* but the shape of Bianliang's walls, a response to natural growth, was contrary to the desired image of the Chinese capital. A plan published in 1933 reflects what the shape of the city might have been: It recognizes the irregular form of the outer wall that was molded in response to demographics and the northwestern position of the palaces (figure 1.7). Other attempts of the 1970s and 1980s to map Bianliang show how hard it is for a Chinese hand

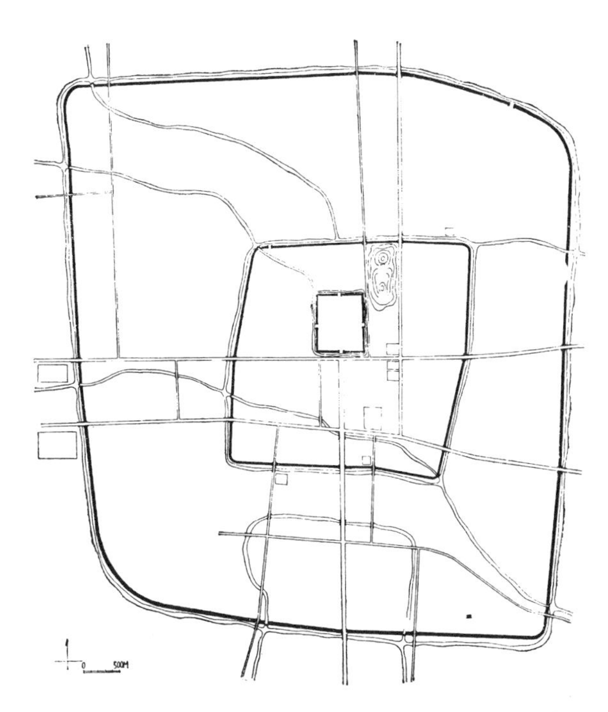

FIGURE 1.8. Plan of Northern Song Bianliang. After Dong et al., *Zhongguo chengshi jianshe shi* [History of Chinese city construction] (Beijing, 1982), p. 43.

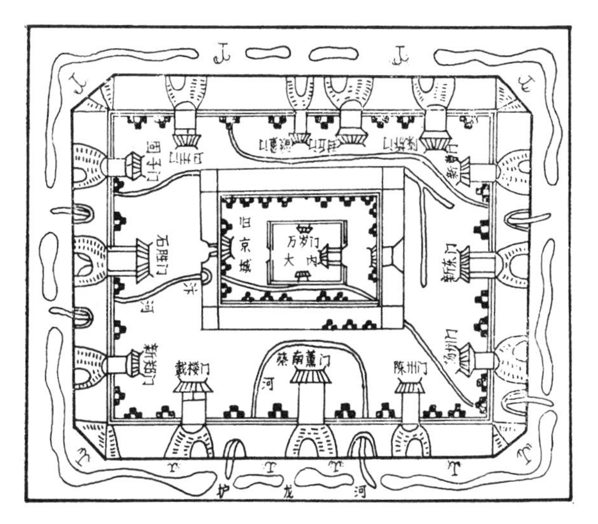

FIGURE 1.9. Idealized plan of Bianliang. From *Shilin guangji* [Compendium of a forest of affairs], 1330s. After Dong et al., *Zhongguo chengshi jianshe shi,* p. 42.

to draw the outer wall of a city as anything but straight or the palaces anywhere but at least roughly centered (figure 1.8).

The explanation for figures 1.6 and 1.8 and for the undue emphasis on the *Kaogong ji* plan in the written history of Chinese urbanism is found in mapmaking at court during the period of Mongolian rule. Specifically, the key to figure 1.6 is a plan of Bianliang like the one published in the 1330s in the court-sponsored encyclopedia *Shilin guangji* (Compendium of a forest of affairs) (figure 1.9). In it one sees not only a source for the truncated corners but, more significant, an obvious derivative of the plan for a ruler's city prescribed in the *Rituals of Zhou*. Indeed, not only was it necessary for Khubilai to have a pedigreed Chinese capital as a symbol of his new role as Chinese emperor, but each successive Mongolian ruler, in true Chinese fashion, continued to legitimate his reign by placing this city in a continuous, if fictitious, lineage of imperial planning that could be traced to the King of Zhou's Wangcheng. Thus more than two hundred years after the Song capital at Bianliang had been destroyed, when no one who had ever actually seen it was still alive, its image had become more powerful than the former reality of any actual city.

Similar examples of the contrast between what topography or records or excavation or a combination of them tell us a city looked like and the way it was presented in officially sponsored Chinese publications after its destruction can be found for two other Song capitals. The Southern Song capital, Lin'an, today Hangzhou, to which the Song regime regrouped after the conquest of the north by the Jin in 1126, is famous for its beauty, which is due partly to irregular topogra-

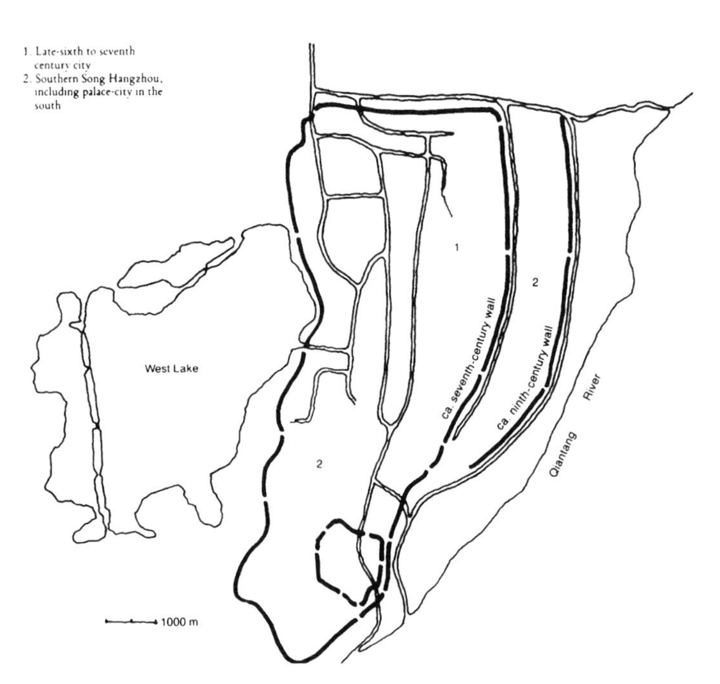

FIGURE 1.10. Lin'an (Hangzhou) in the Southern Song. After Steinhardt, *Chinese Imperial City Planning,* p. 25. Courtesy of University of Hawaii Press.

phy. The city of Hangzhou winds around what is arguably China's most beautiful body of water, West Lake (figure 1.10).

Yet an official publication of the period 1265–74 illustrated the plan of the city as a central palace area inside a nearly square outer wall with two cut-off corners (figure 1.11). Thus, even as China's ultimate place of beauty was being barbarized by the Mongols, maps of the city printed under Chinese court sponsorship were subject to the same practices undertaken in Mongolian court publications. In thirteenth-century China, an officially published map could thus guarantee that, whatever happened to the timber, brick, tile, and stone, the image preserved for posterity was of a city as Chinese, as pedigreed, as legitimate, and as orthodox as the city prescribed for the king of Zhou.

One of the most amazing plans of a Song capital is one labeled, simply, "western capital" (figure 1.12). Its source is *Henan zhi* (Record of Henan province), a Mongolian period record preserved in the 1408 encyclopedia *Yongle dadian,* source of the earliest extant map of the king of Zhou's city. The city illustrated is Luoyang, western capital of China beginning in the tenth century. What Luoyang looked like up to the end of the ninth century has been made known through excavation. The secondary capital of China for most of the seventh, eighth, and ninth centuries, Luoyang had been for three hundred years of Tang history a city modeled after Sui-Tang Chang'an (see figure 1.4). It is highly unlikely that the movement of all three walls occurred in the tenth century. Yet the Yuan period compiler of the court record could assume, it seems, that someone looking at this plan would not necessarily seek additional literary records or verification of what survived at the city. What a city had looked like in the past was less relevant, even irrelevant, in comparison to the purpose of printed words or pictures in books, the unquestionable authorities and true transmitters of past China to future China. In a sense even Marco Polo was part of the literary transmission of misinformed details about Chinese imperial cities based on the widespread dogma of what a Chinese city was supposed to look like. For whatever reason, the outer wall of Dadu had only eleven gates (see figure 1.5). Marco Polo wrote that it had twelve (number of the king of Zhou's city), leaving one to ponder if in all his years in the Chinese capital the Italian merchant ever walked along the northern city wall.[12] Yet perhaps the greatest irony is the determination to maintain the importance of a city center even through the iconoclasm represented by other actions of the redesign of Beijing after October 1, 1949. Mao Zedong proclaimed his new government from

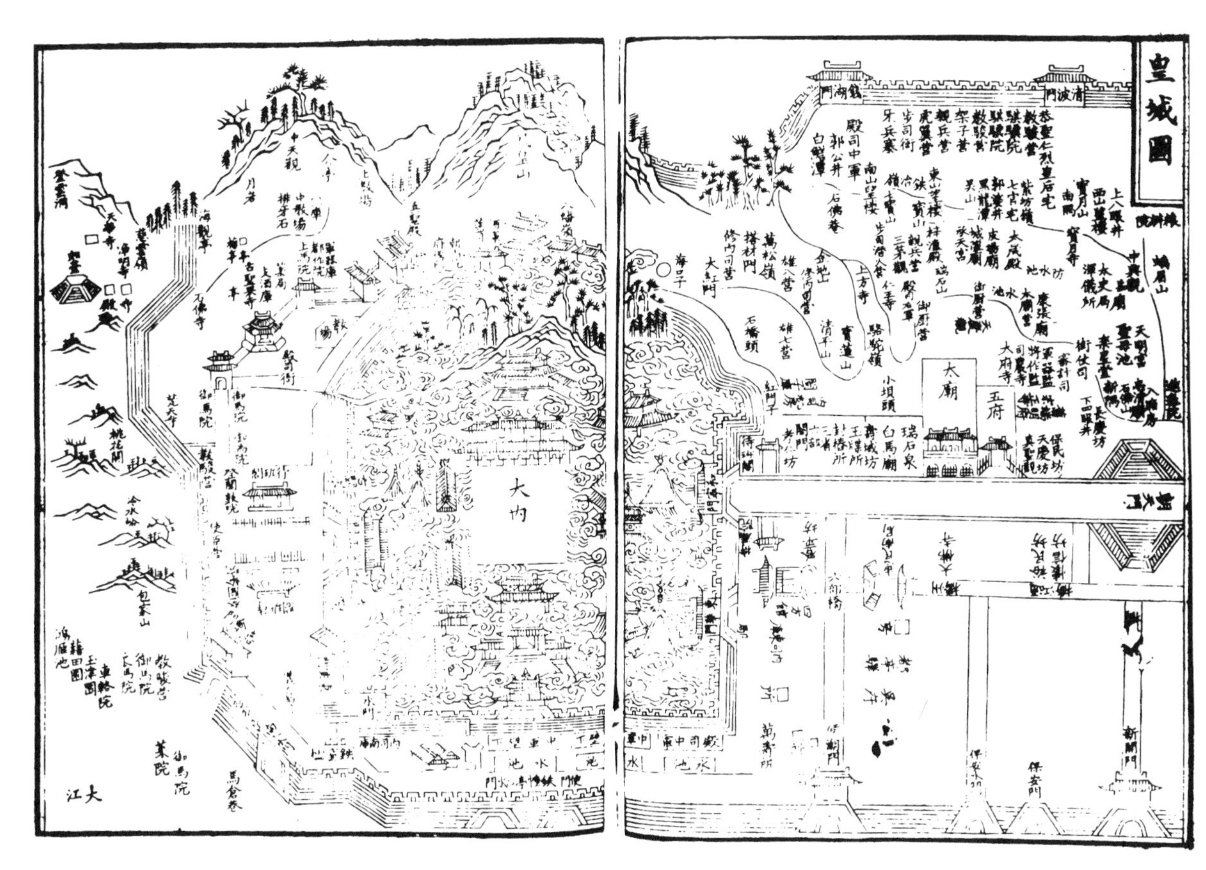

FIGURE 1.11. *Huangcheng tu* [The imperial city (Lin'an)]. From *Xianshun Lin'an zhi,* 1867 rev. ed. of thirteenth-century record.

near the spot where he is entombed, the focus of the centuries-old north-south axis of Beijing where former Chinese emperors had sat facing south in their Wangcheng-inspired cities.

Broader Issues of the Text, the Image, and the Reality

THE DESIRE TO ENVISION, if not to map, a ruler's city as ideal, even if this understanding requires mythology, fantasy, or forgery, is somewhat universal, especially among ancient civilizations. One can begin with comparisons between Chinese and imperial Roman cities. In *The Idea of a Town,* Joseph Rykwert discusses proto-Rome as an example of a "built symbolic pattern" evident of "rectilineal planning and orientation" (to cardinal points).[13] One can go farther in comparisons of the idealized city of the king of Zhou and the town founded by Romulus to discuss the important role of the planner, the overriding concerns of myth and ritual as opposed to economics or hygiene in the founding of a city, and the entombment of the founder in the center of the city (the last a place occupied by Mao alone among the rulers of China).[14] Further, one can cite Aristotle's recommendation that a site be oriented to the south[15] and juxtapose this thought with the universal orientation of Chinese imperial cities

西京城圖

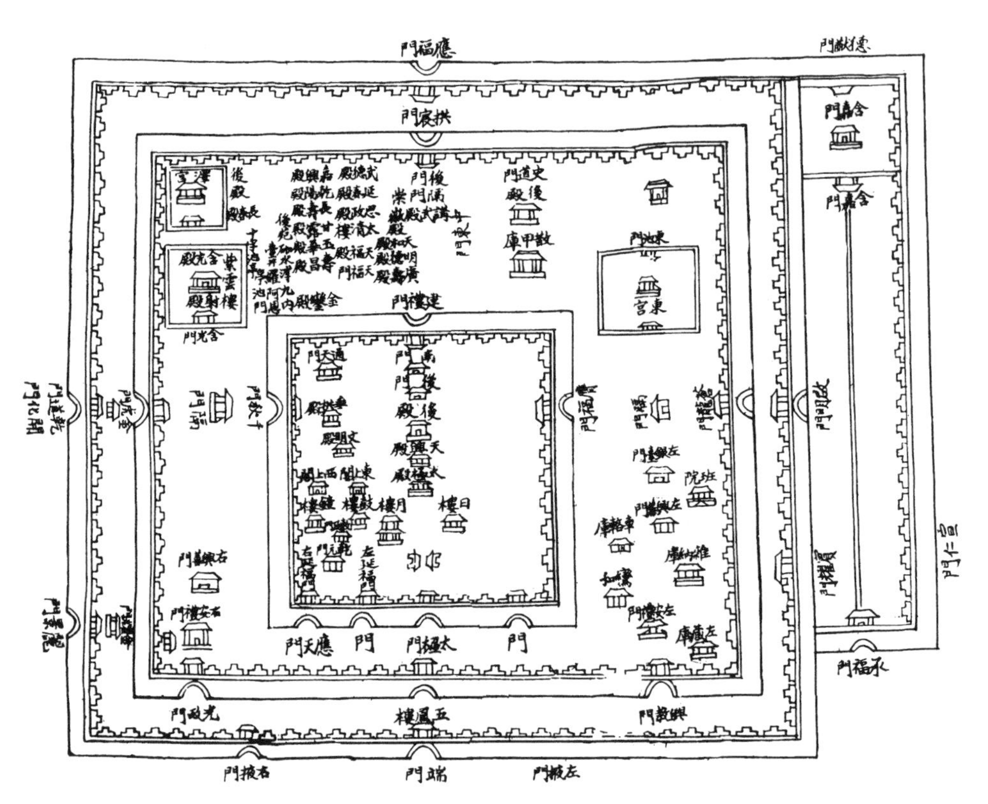

FIGURE 1.12. *Western Capital* [Northern Song Luoyang]. From *Henan zhi* (Yuan period) as preserved in *Yongle dadian, juan* 9561 (Taibei reprint, 1974), fifth plan.

southward. A fascinating Sino-Roman comparison involves the issue of divination. Although the medium of divination was different, both the Chinese and the Romans believed that site determination could influence the fate of a people and that divination was a means of ascertaining correct siting. In ancient Rome the medium was a liver and the process haruspex. In China the medium was the bone of a cow, pig, or sheep or a tortoise shell and the process scapulimancy or plastromancy.[16] Yet another comparison can be made between Pliny's directive that the cast of a shadow be used to determine the points of orientation[17] and the *Kaogong ji* passage quoted above suggesting the cast of a shadow as the determinator of a midpoint. Finally one comes to the fascinating Sino-Roman comparison involving bronze cauldrons. It has been suggested that eight bronze amphorae discovered in a shrine in Paestum may have been associated with the hero-founder of the city.[18] According to Chinese legend nine bronze vessels (*jiu ding*) were cast during the time of the legendary emperor Yu (in the third millennium B.C.). These bronzes were the possession of every Chinese emperor and as a group they were a symbol of his legitimate rulership.[19]

What is especially interesting, however, is a relationship between the bronze symbols of imperial rule and one of the earliest recorded maps of Chinese territory. In relating an incident of a military attack in the year 605 B.C., the circa third-century text *Zuo zhuan* (Commentary of Zuo) makes mention of the nine tripods cast by the legendary emperor Yu the Great. On them the ruler is said to have cast "pictures of things natural to distant regions." This passage may be interpreted as the earliest textual reference to Chinese maps.[20] Further implied is that from earliest times mapping was an imperial activity in China. The casting of maps on bronze, the most permanent material of the age, elevates cartography to an immutable plane for which the actual depiction is as important as the association.

Beginnings of the Chinese Tradition

JUST THE ONE literary passage suggests that mapping of cities or regions in China was from earliest times permanent in a way that urban cartography was not in Hellenistic times nor would become in the West. Indeed, comparisons between Chinese and ancient Roman mapping break down because maps designed in accordance with the description in the *Rituals of Zhou* were published in China more than seventeen hundred years after the time of Confucius in spite of changes in urban space that occurred in the intervening centuries. Thus even if excavation were to confirm that ancient Roman cities did not so closely resemble the published orthogonal plans and literary passages that extol such planned or described cities, such as is the case for China, after the fall of the Roman Empire irrefutable changes would occur not only in Western cities but in pictures of them.

The development of Chinese urban cartography is not what one might have expected, given the sophisticated levels of mapmaking and literature about mapmaking that existed in China in the first millennium B.C. The oldest known Chinese site plan survives on a bronze plate inlaid with gold and silver (dimensions 94 × 48 × 1 cm) uncovered in the 1970s in a tomb of a Zhongshan king named Cuo in modern Pingshan county, Hebei (figure 1.13).[21] Grave goods and external evidence suggest that burial took place in the second to last decade of the fourth century B.C. Shown on the bronze plate are five sacrificial halls and four smaller buildings north of them (south is at the top), two enclosing walls, and a mound of about fifteen meters on which all the structures were elevated. This earliest example of a bird's-eye view in China was also drawn to scale. The innermost enclosed area is approximately 1/500 of the original. Dimensions of the buildings and distances between them are given in *chi* and *bu*, one *chi* being about six *bu*. Those for whom the sacrificial halls were intended are also known: one was for the king, two for queens, and two for lesser wives. Only two of the intended five occupants of the necropolis are interred there. Three of the group outlived the fall of the kingdom. Again, we have evidence of a totally planned space the reality of which was different from the permanent image. A forty-two-character inscription on the plate gives pertinent information about the earliest mapping of Chinese cities: Two copies of the map were made, one for burial and the second intended to remain in the palace. The palace version suggests the importance of maps as documents of past activities of royalty. It further may suggest an anticipated reuse of the plan in future generations.

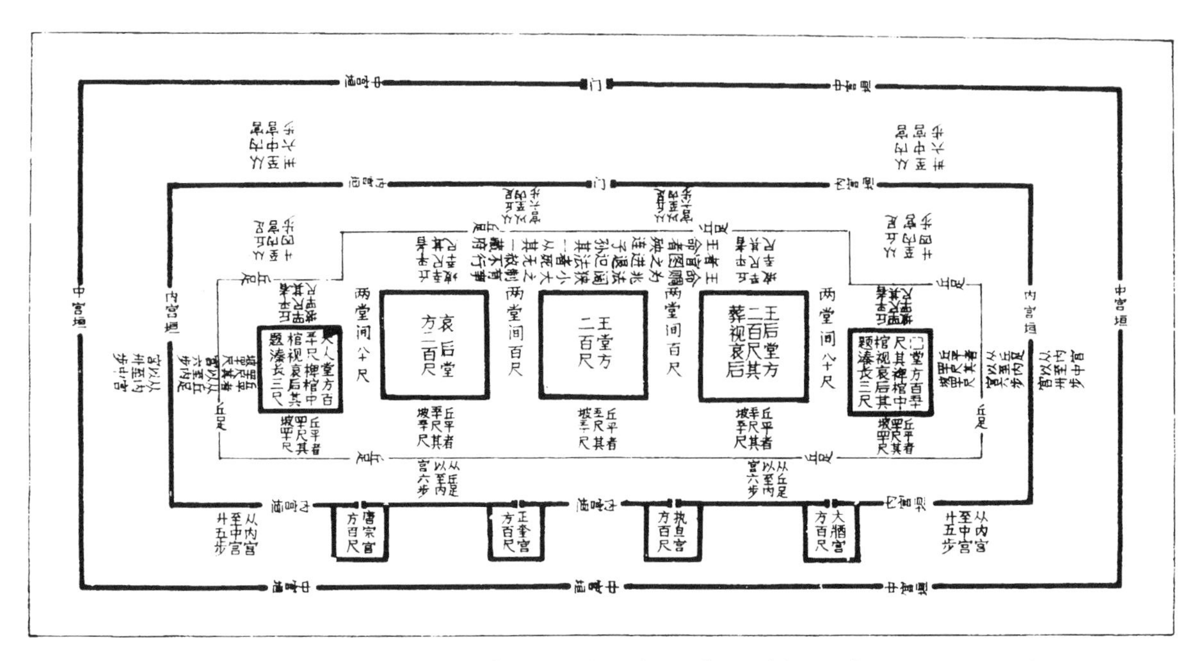

FIGURE 1.13. Line drawing of *Zhaoyu tu* with inscriptions in modern Chinese characters. From Cao Yanru et al., *Zhongguo gudai ditu ji* [Atlas of ancient Chinese maps], (Beijing, 1990), 1: pl. 3.

Seven maps drawn in ink on one or both sides of pine boards were excavated from a tomb dated 239 B.C. at Fangmatan, near Tianshui, Gansu province.[22] All the maps depict regions much larger than sites or cities in the vicinity of the Wei River near the Gansu-Shaanxi border. It has been suggested that the maps may have been used by military guards or in the timber-logging operations in the Tianshui region.[23] In a later tomb (179–41 B.C.) from Fangmatan a paper map was uncovered. Although too fragmentary for identification of its picture, it is now the oldest example of Chinese paper.

The best known early Chinese maps on silk were excavated at the tomb of another Chinese official, the son of the marquis of Dai who was buried in 168 B.C. at Mawangdui, Hunan province.[24] Both a topographic map and a military map, drawn to scale and employing symbols for mountains, streams, roads, towns, and a temple, have been reconstructed.[25] Most significant for this study was the city map uncovered in the tomb (figure 1.14). Painted on the same piece of silk as exercise techniques, the city map shows elevations of buildings situated on orthogonally laid out streets. One cannot help but notice the similarities between the arrangement of streets and the street plan described in the *Kaogong ji* and shown in illustrated versions of that text.

No one familiar with Chinese civilization is surprised that records on a fourth-century B.C. plan are corroborated by textual sources. Perhaps more impressive is the degree of instructive accuracy on a bronze plate that should have made it possible for builders to copy this plan as a five hundred times larger, three-dimensional building group. Equally noteworthy is the symbolic key worked out by the second century B.C. All

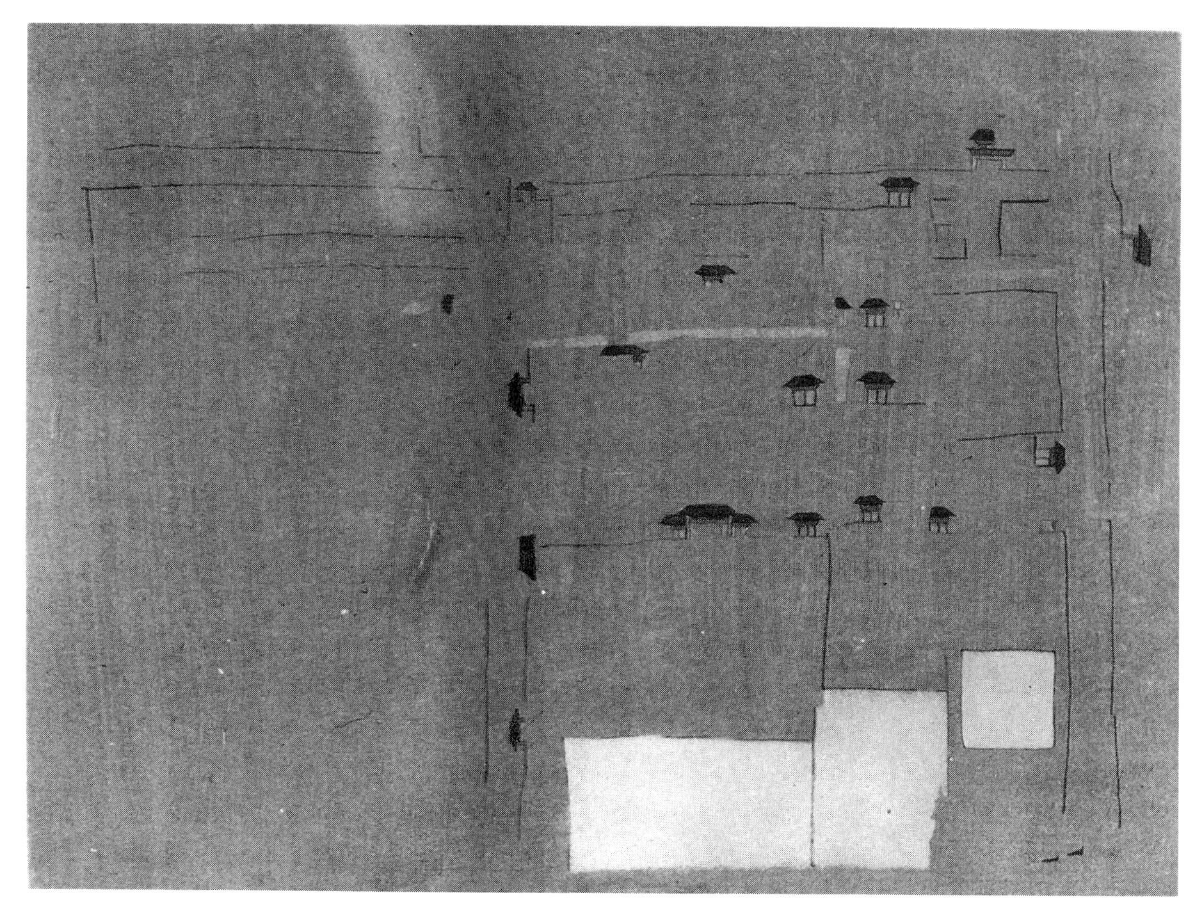

FIGURE 1.14. Copy of city or county map excavated at Tomb 3, Mawangdui, Hunan, ca. 168 B.C. From Cao Yanru et al., *Zhongguo gudai ditu ji,* 1: pl. 29.

the early maps suggest reading of both lines and words, implying a standard of literacy among at least some military personnel and builders in ancient China. Clearly evident, too, in the early Chinese maps is the close relation between the written word and the plan. The association forged in pre-Han maps would, like the image of a ruler's city described in the *Rituals of Zhou,* be maintained through the rest of China's imperial history.

"Written maps," or unillustrated texts that serve the same purpose as maps, were made during the Warring States period (ca. 781–477 B.C.), even before the casting of King Cuo's *zhaoyu tu.* The earliest Chinese term for map, *ditu* (picture of the earth), is from the same period. Still the most common word in colloquial Chinese for map, it is also found in the fourth- and third-century B.C. literary work *Guanzi* in which one of the purposes of early mapmaking is made clear. "All military commanders must first examine and come to know maps" is the opening line of the section devoted to *ditu* in the classical text.[26]

Three literary works, all first written before

the fourth century A.D., and all originally published with accompanying maps, can be considered part of a genre of writing known as *tu jing*, classical works with illustrations. The *Shanhai jing* (Classic of mountains and waterways), written about the time of the Han dynasty, was originally accompanied by a map of China that has not survived. *Yu gong* (Tribute of Yu), part of the *Shu jing* (Book of history), from about the sixth or fifth century B.C., also originally had a map of China as did the geography section (*dili zhi*) of the first-century A.D. *Han shu* (History of the Han dynasty). *Dili zhi* became a standard section in subsequent dynastic histories, and provincial or more local records (*difang zhi*) also frequently would include maps.[27]

Of particular relevance to mapping the Chinese city is a third-century A.D. work entitled *Shui jing* (Water classic), which survives as a major early fifth-century study of Chinese geography because at that time Li Daoyuan wrote an extensive commentary on it. Accompanying maps for the *Shui jing*, including maps of cities, were published in the nineteenth century by Yang Shoujing. Among Yang's maps is one of Pingcheng, the fifth-century capital near today's Datong in Shanxi (figure 1.15). No evidence can confirm the arched walls or central palace city. More probable is that the city resembled earlier or contemporary cities of North China built by the Northern Wei like Ye or Luoyang of the third to sixth centuries that had palace cities in the north center. The plan of Pingcheng is late evidence of the practice employed at the Mongolian court described above whereby idealized plans, especially plans that favored the description of the king of Zhou's Wangcheng, were substituted for cities whose physical remains either did not resemble that shape or could not be proved to have resembled it.

By the period of Pingcheng's ascendance to capital status in the fifth century, mapmaking in Chinese society was well established in two main roles. The first, the military role, produced maps of increasing accuracy over time. Yet by nature the plans were restricted and never accessible outside the court. Maps that were not made for military purposes seem nevertheless to have circulated among an equally self-contained sector of Chinese society, the educated elite. Yet, whereas maps for military purposes tended to be of the analytical tradition of cartography, the second type of maps, accompaniment to scholarly

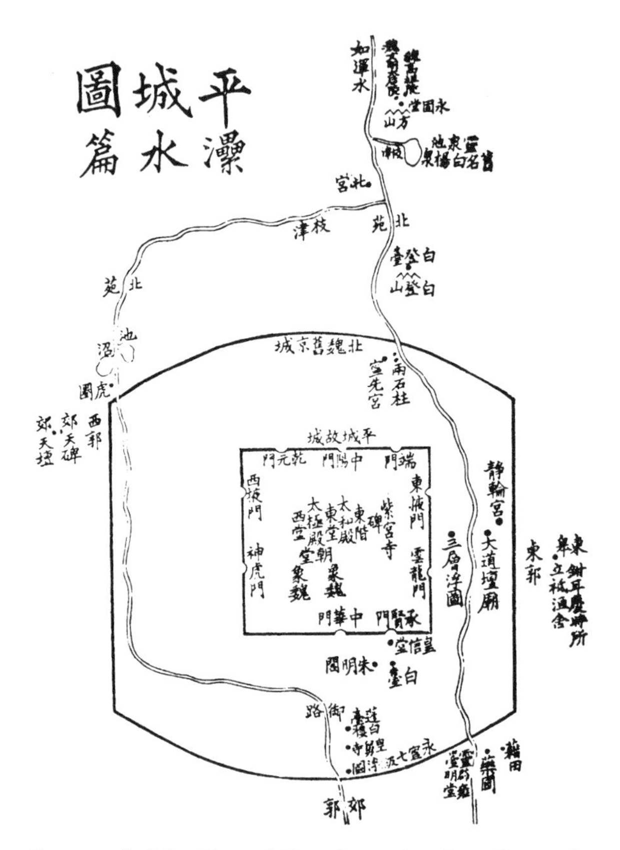

FIGURE 1.15. Plan of Pingcheng (modern Datong), in the Northern Wei. From Yang Shoujing (1839–1915), *Shui jing zhu tu* [Illustrated commentary on the *Water Classic*] (Taibei reprint, 1966), p. 77b.

FIGURE 1.16. *Panshan tu* [Picture of Panshan]. From *Jizhou zhi* [Record of Ji prefecture], vol. 1 (*Hebei fangzhi*, vol. 6), pp. 46–47, 1968 reprint of 1831 original.

writings, may be thought of as descriptive cartography.[28]

This second group of maps often was composed of actual pictures of places. Historically, Chinese of the elite class traveled through their cities and described them. Yet it was not necessary for a scholar-official to have seen a place in order to write about it, or for its map to accompany scholarly compilations or writings. Often maps produced by China's educated elite for nonmilitary purposes are found in local records (*difang zhi*) published by provincial, prefectural, or even more local government offices. Figure 1.16, "Picture of Panshan" (in Hebei province), is a typical example of the descriptive cartography produced by this group.

From almost as early as there are records of the activities of Chinese elite, one finds that when cartography was not for military purposes it was a pastime of the educated elite or a backdrop for court pleasures. Maps were used, for instance, as background for human-animal combats at the court of Han emperor Wudi (r. 140–87 B.C.).[29] An incident recorded in the mid-ninth-century *Lidai minghua ji* (Record of painters of successive generations) relates that when Sun Chuan, ruler of the Wu state, commissioned a map of his territory in the third century A.D., painters were called on to make it. His wife, also a skilled painter, embroidered a map for him.[30]

It is not surprising that third-century Chinese court painters made maps, for strikingly parallel developments can be observed in the histories of painting and cartography in China during the period of the Northern and Southern Dynasties (ca. third through sixth centuries). Through these similarities the second purpose of maps is even clearer. The first parallel is that neither mapmaking nor picture making had broken away from the written word, nor, by extension, from an official and often didactic purpose. Chinese painting would eventually flourish independent of court sponsorship whereas cartography would not, but in the period of the Northern and Southern Dynasties painters were still titled and salaried by the court and their "art" served the court either as records of events or as pictures that were seen as illustrative appendages to the texts they accompanied.[31] The same can be said of cartography. More fascinating, during the period of Northern and Southern Dynasties, Six Laws were articulated for both painting and cartography. In his preface to *Yugong diyu tu* (Maps of the regions of *Yugong*), the only surviving part of the text, government official Pei Xiu (224–71) promulgated Six Laws of Mapmaking. The two-character, pithy rules have been translated as: (1) graduated divisions (to determine scale); (2) rectangular grid (for determining relative position of one place to another); (3) pacing the sides of right triangles to determine the distance traveled; (4) leveling of high and low; (5) measuring right and acute angles; and (6) measuring curves and straight lines.[32]

The purpose here is neither to argue with published translations of these principles (see n. 32) nor to try to determine (as those who have translated the principles have done) the extent to which they might have been put into practice. Simply, it is significant that both their two-character structure and the age in which they were promulgated are the same as those of the Six Principles (of painting) expressed by Xie He.[33] Both sets of rules were part of the general knowledge of every later practitioner of the trade.

Aspects of the better studied role of painting in third through sixth century Chinese elite society and its culture aid in our understanding of the role of cartography. Chinese maps of cities that were not strictly guarded military property were produced by the same elite group who painted as well as cultivated the arts of poetry, calligraphy, music, and game playing. Possessing artistic status, the mapper was encouraged to take artistic license. In the case of the Chinese map, this does not mean decoration but license to make the map that best serves a purpose. Just as famous paintings of intrinsic historical value were copied at the Chinese court, one can imagine old maps were available as models in national, provincial, or prefectural offices. For maps, however, the choice of old models resulted in oblivion to modern techniques, ultimately separating Chinese city maps from non-Chinese ones.

Systems of Accuracy

Certain kinds of accuracy were nevertheless achieved even in very early Chinese maps of cities. Relative accuracy of place to place, for instance, is found in Han maps that survive in the form of murals. A variety of maps are painted on the walls of a tomb uncovered at Helinge'er in the Inner Mongolian Autonomous Region.[34] Although not painted to scale, one finds, in the painting of the owner's manor, spaces for different purposes and their relative positions to one another (figure 1.17). In contrast to the representation of these more private spaces of the estate, in the depiction of public space (the town where the tomb occupant served as an official [figure 1.18]) cardinal orientation is as clear as it is in early Chinese plans such as the necropolis of the Zhongshan kings or the painting on silk

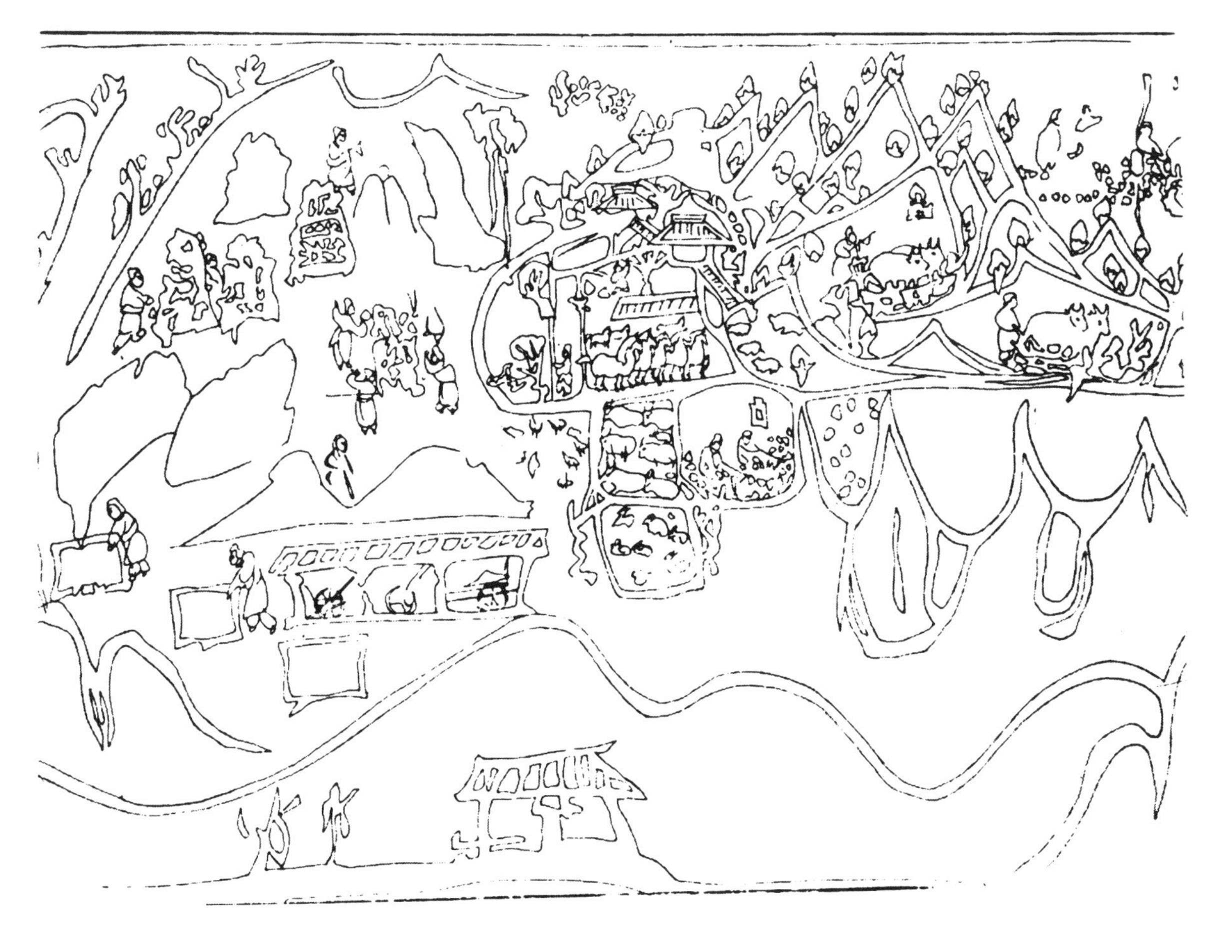

FIGURE 1.17. Line drawing of tomb occupant's manor. From Han tomb, *Helinge'er Hanmu binua* [A wall-painted tomb from the Han dynasty at Helinge'er], Beijing, p. 21.

from Mawangdui Tomb 3 (see figures 1.13 and 1.14). In other words, the principles of urban cartography differ according to function even in the same tomb—orthogonal lines are used to depict official space and more calligraphic brush strokes delineating irregularly shaped areas represent private, residential space.

During the Tang dynasty (618–906), distinctions apparent in the Han tomb maps of the countryside in comparison to official or imperial space are still found in wall painting. Representations of imperial space on the walls of the tomb of crown prince Yide from the first decade of the eighth century show an exceptional kind of descriptive accuracy of place to place. Since excavation and publication in the early 1970s of the tomb of Tang crown prince Yide, one has suspected that the gate towers at the entrance to his "subterranean palace" (*xia gong*) represented the gate towers of Chang'an, the city he would have ruled had he not been executed (figure 1.19).[35] These pictures were used immediately after they were discovered to inspire theoretical reconstructions of palatial halls and gates that had

FIGURE 1.18. Map of Fangyangxian (county). From Han tomb, Helinge'er, Inner Mongolia, second century A.D. Cao et al., *Zhongguo gudai ditu ji,* 1: pl. 32.

FIGURE 1.19. Gate towers. Underground approach path to tomb of Crown Prince Yide, Xi'an, ca. 710. From *Han-Tang bihua* [Wall painting Han to Tang], (Beijing, 1974), pl. 86.

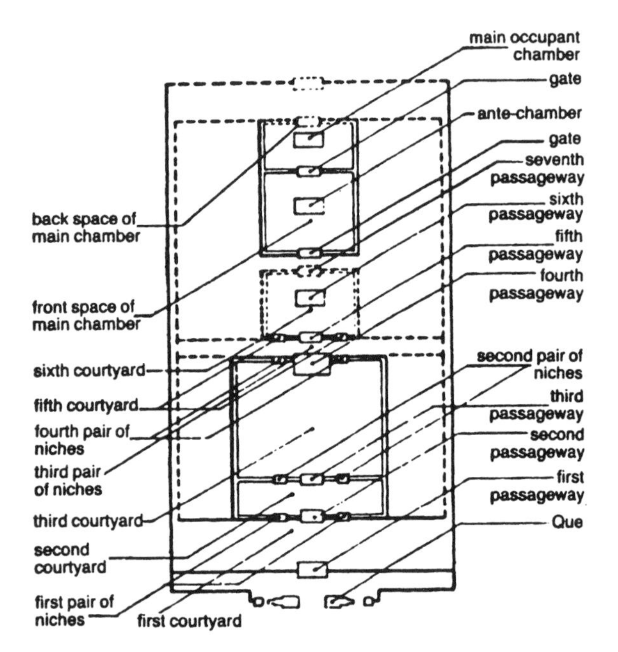

FIGURE 1.20a. Plan of buildings shown in wall paintings of approach to tomb of Crown Prince Yide. After Fu Xinian, *Wenwu yu kaogu lunji,* p. 340. Courtesy of University of Hawaii Press, Steinhardt, *Chinese Imperial City Planning,* p. 108.

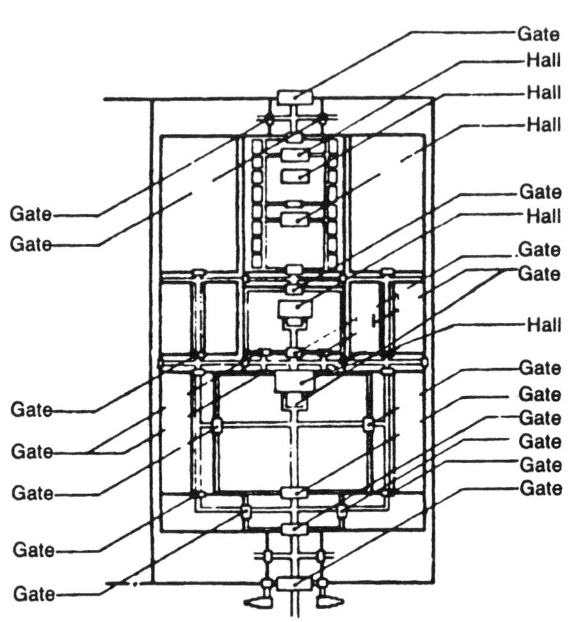

FIGURE 1.20b. Reconstructed plan of palace complex of Crown Prince Yide based on texts. After Fu Xinian, *Wenwu yu kaogu lunji,* p. 340. Courtesy of University of Hawaii Press, Steinhardt, *Chinese Imperial City Planning,* p. 108.

been uncovered at the Tang capital about fifteen years earlier.[36] In 1987 a specific correspondence was realized between the wall images (shown as a plan in figure 1.20a) and the plan of the eastern palace where the Tang crown prince resided in life as it is described in texts (figure 1.20b).[37] This private painting sealed in an eighth-century imperial tomb is comparable in accuracy to that of the bronze plate excavated at the fourth-century B.C. tomb in Pingshan, but the accuracy of scale in the wall painting cannot be verified.

The same architectural historian who drew figures 1.20a and 1.20b published a study of the wall paintings in the twelfth-century Mañjuśri Hall of Yanshan Monastery in Fanshi, Shanxi (figure 1.21).[38] In this second article he transferred the background architecture for the life of a Buddhist deity into a plan (figure 1.22a) that was startling similar to the arrangement of the twelfth-century palace city at the contemporary central capital as it is described in a text (figure 1.22b). Clearly, Chinese painters, like cartographers, since the Han dynasty were capable of certain kinds of accurate detail. The accuracy of the paintings of architecture on the walls of the early-eighth-century tomb and the twelfth-century temple walls may even suggest that the paintings were based on line-drawn plans. However, such plans would not have had wide circulation. Their audiences were much more limited, in the case of the royal tomb paintings, probably restricted, in comparison to published maps of cities or smaller walled spaces.

Wall painted images of nonimperial Tang

FIGURE 1.21. Line drawing of paintings from western wall of Yanshan Monastery South Hall, Fanshi, Shanxi, ca. 1167. After Fu Xinian, *Jianzhu lishi yanjiu* 1 (1982), between pp. 130 and 131.

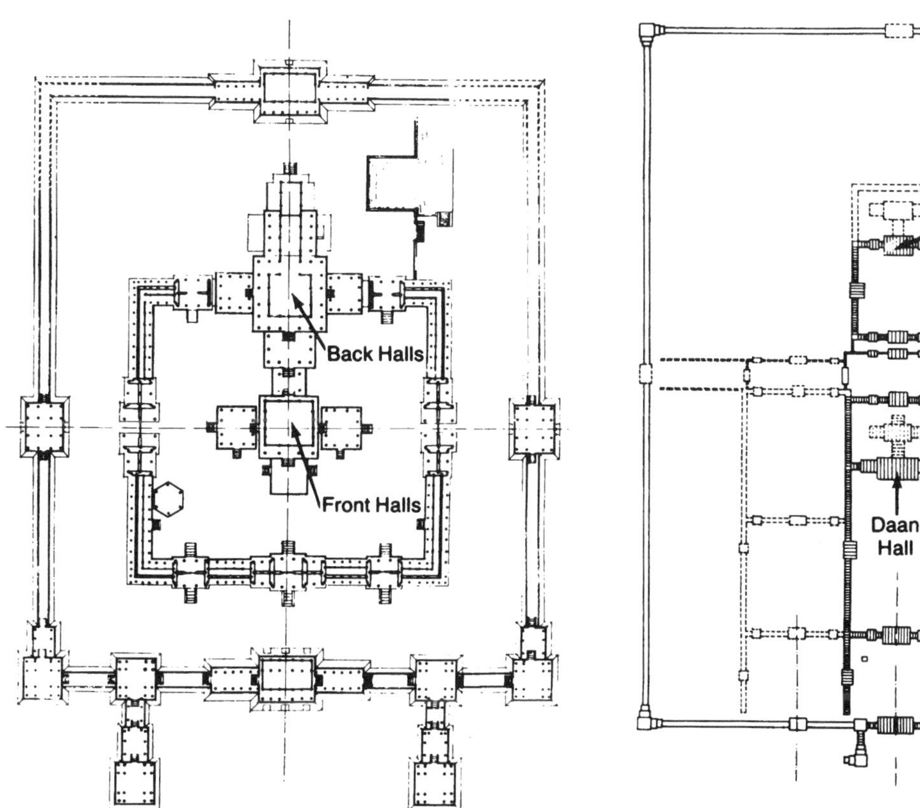

FIGURE 1.22a. Reconstructed plan of figure 21. After Fu Xinian, *Jianzhu lishi yanjiu* 1 (1982), p. 132.

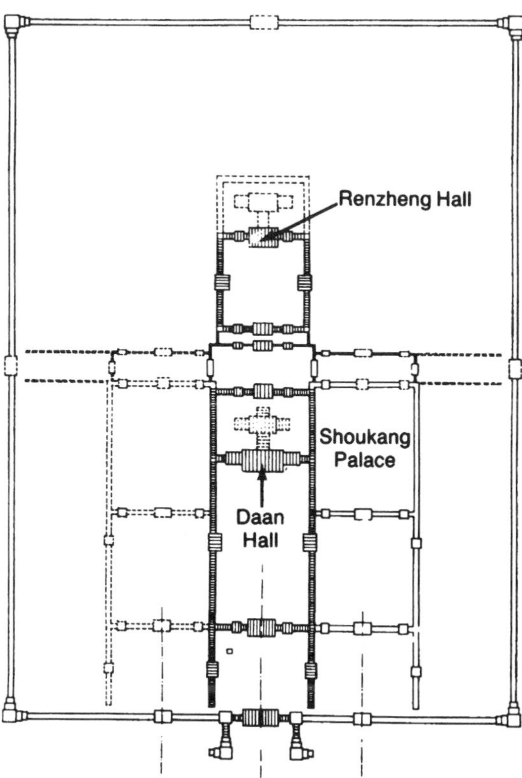

FIGURE 1.22b. Plan of palace city of Jin dynasty central capital of twelfth century after *Lanpei lu*. After Fu Xinian, *Jianzhu lishi yanjiu* 1 (1982), p. 133.

FIGURE 1.23. Left side of painting from western wall of Cave 61, Mogao Caves, Dunhuang, tenth century. After Cao Yanru et al., *Zhongguo gudai ditu ji,* pl. 40.

space executed outside the court are, like the countryside home of a Han official and most local-record drawings of cities, descriptive cartography. Outstanding examples survive in Cave 61 of the Mogao Caves near Dunhuang, Gansu province. The tenth-century wall paintings show eight towns and more than sixty pilgrimage monasteries, including pilgrimage stops, between the Foguang Monastery in Wutai county, Shanxi, and Hebei province (figure 1.23).[39] The painting probably functioned not so much as a map for directions but rather as an "after the fact" picture. While worshiping in the cave a pilgrim could pause to look at the wall painting and say, "Yes, I was there," or, "Next I would like to go there." For actual travel purposes he probably would have relied on word-of-mouth directions en route.[40]

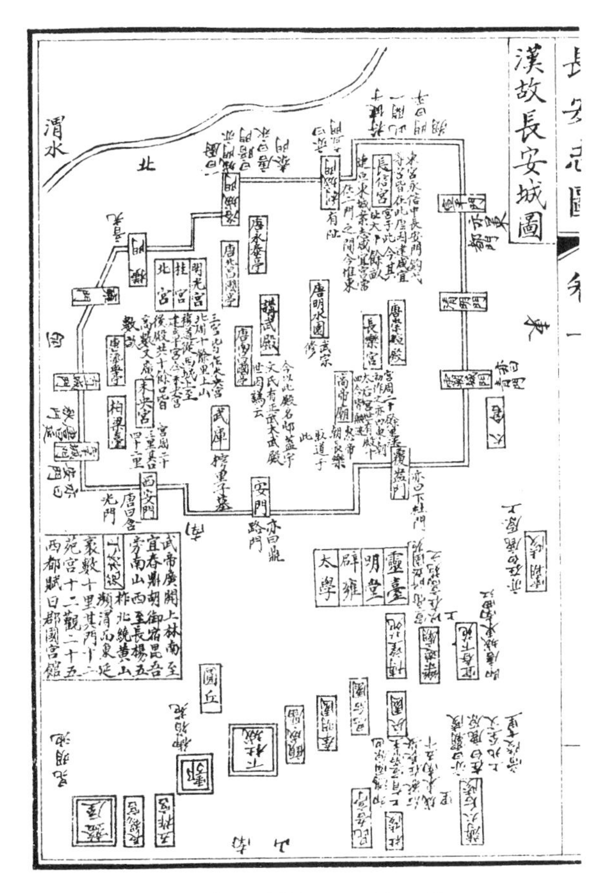

FIGURE 1.24. Plan of Western Han Chang'an. From Li Haowen (Yuan dynasty), *Chang'an zhi tu* [Illustrated record of Chang'an], reprinted in Bi Yuan (1730–1799), ed., *Jingdiao Tang congshu,* no. 24, *juan* 1/5.

Thus Chinese mapmakers were not universally limited to orthogonal lines, but the majority of preserved maps of cities have them because so much of the mapping of Chinese cities was tied to court or imperial sponsorship or purposes. Thereby a scholar at court could draw a plan of a city or city space he had never seen and be convinced that its accuracy was unquestionable, at least to the extent that accuracy might have been a concern. According to the standards of mapmaking at the Chinese court, such a map was accurate.

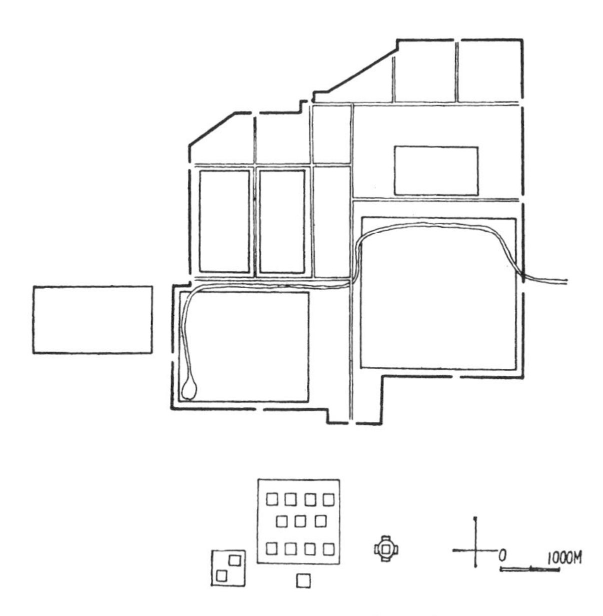

FIGURE 1.25. Postexcavation plan of Western Han Chang'an. After Dong Jianhong, *Zhongguo chengshi jianshe shi,* p. 17.

HAN CHANG'AN AND ITS EXCEPTIONAL PLANS

AMID THE RULES, restrictions, and resulting designs, maps of one Chinese city are exceptional in their recognition of the unusual shape of the city's outer wall. Yet even they exemplify the purposes of urban cartography in China.

The true form of the outer wall of Chang'an, in particular its jagged northern and southern faces, was recognized (figure 1.24) long before it was confirmed in postexcavation maps (figure 1.25). In fact, in contrast to the other Mongolian period maps of Chinese capitals discussed here (see figures 9, 11, and 12), the similarities between figure 24 and the archaeologically generated map are striking.

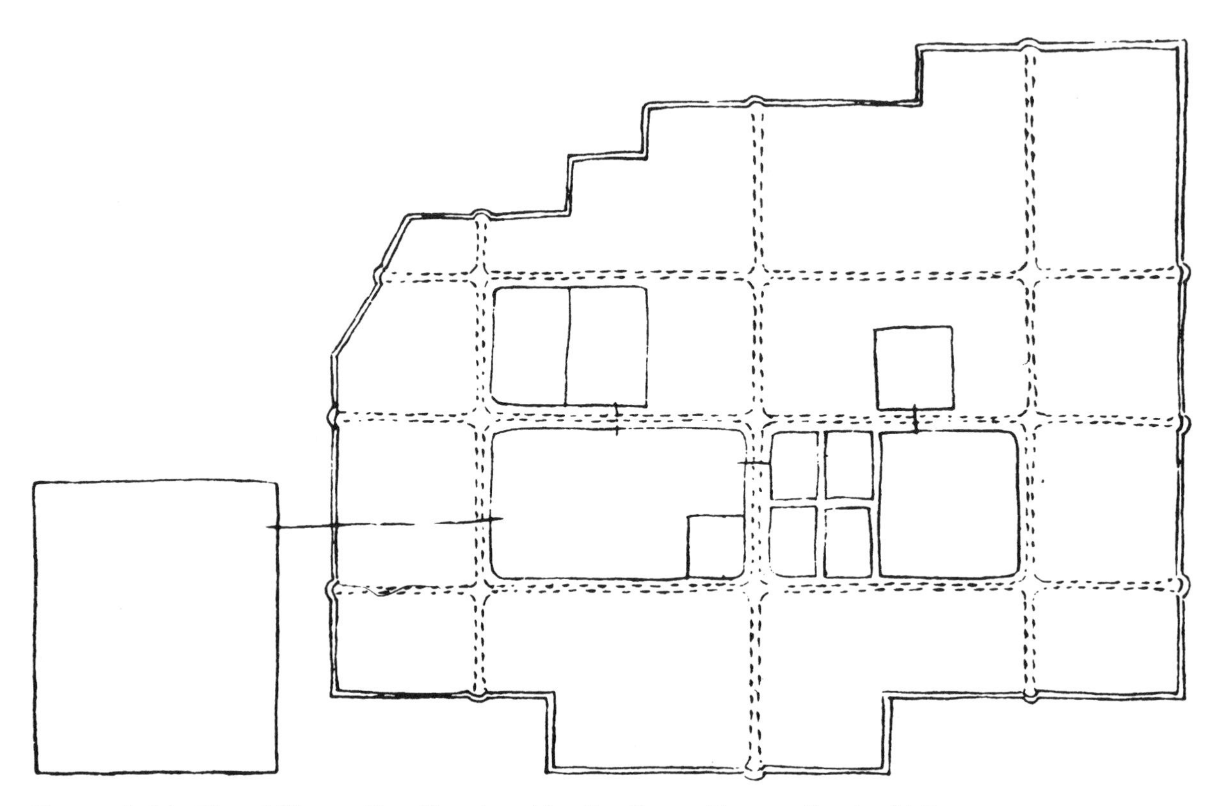

FIGURE 1.26. Plan of Western Han Chang'an. After Yao Jiazao, *Zhongguo jianzhu shi,* fig. 102.

Han Chinese sought to explain their city scheme by cosmology: the *Shi ji* (Record of history) and *Sanfuhuang tu* (Illustrated record of the three districts of the capital), for instance, relate that the city was constructed to conform with the Great and Little Dippers.[41] In actuality, the unusual configuration can be explained by practical considerations. First, the Han rulers followed Qin precedent in the construction of their city: palaces were rebuilt on the ruins of Qin building complexes. Second, the imperial city of Qin-Han was essentially a set of walled palaces, designed primarily as protection for the ruler. The wall was shaped to enclose those preexistent structures. Furthermore, the unusual shape of the Han wall in the north of the city conformed to the flow of water, vital for transport and protection. Yet a plan of Han Chang'an published in the 1930s suggests that, in spite of the irregular outer-wall shape, the interior streets conformed to the *Kaogong ji:* Palace compounds are fit into a scheme of three north-south by three east-west streets (figure 1.26).

CARTOGRAPHY, CALLIGRAPHY, AND THE GRID SURFACE

ANOTHER PLAN OF THE WESTERN HAN capital shows that even when the three-by-three street scheme is not implemented, it can seem implicit due to the grid surface on which the map is drawn (figure 1.27). In fact, an appeal of the *Kaogong ji* text was probably that it could be visual-

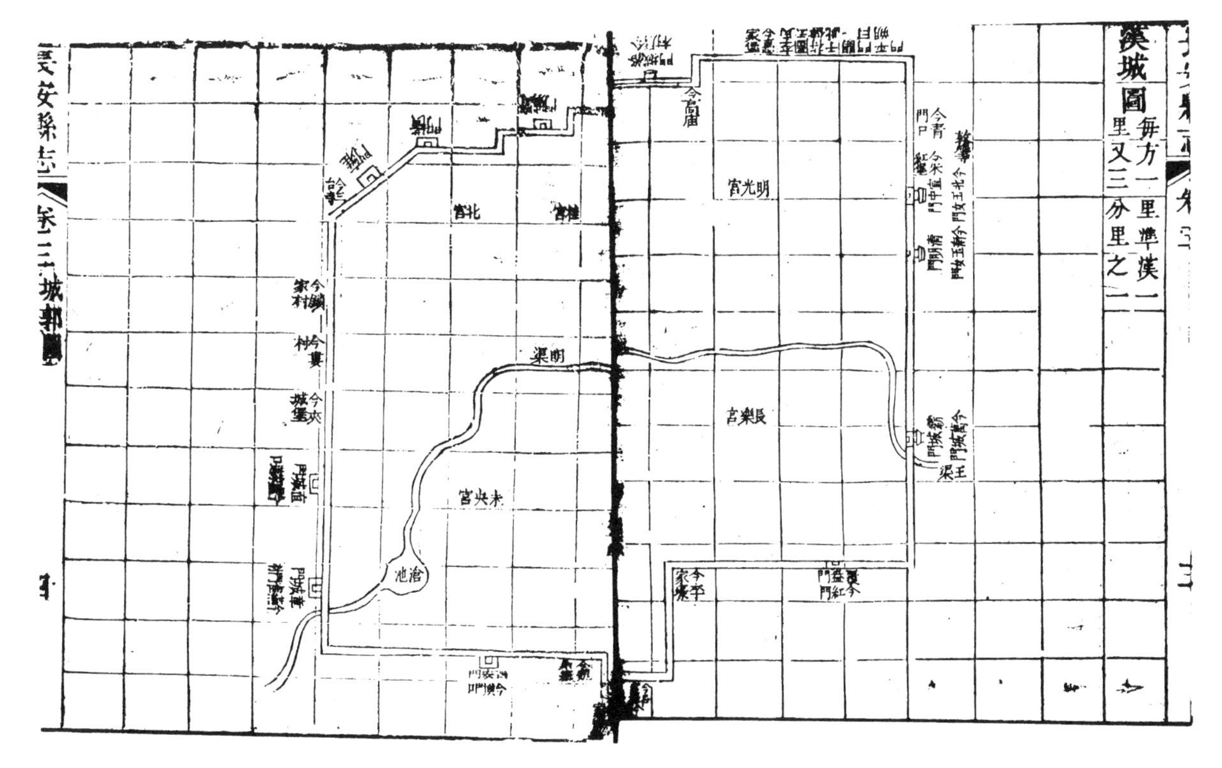

Figure 1.27. *The Han City.* From *Chang'an xian zhi* [Record of Chang'an district], 1815, *juan* 3/3b–4a.

ized in two dimensions as a grid. Local records of pre-twentieth-century towns and prefectures provide evidence that when a city or territory outline was more irregular even than Han Chang'an's, it still might be mapped on a grid (figure 1.28). The visual suggestion is that evenly shaped enclosures somehow exist not just within all outer city walls—even though only Tang Chang'an (see figure 1.4) and perhaps fifth-century Luoyang and Ye were truly divided into districts as regular in shape as those on a grid—but beyond the walls into the Chinese countryside.

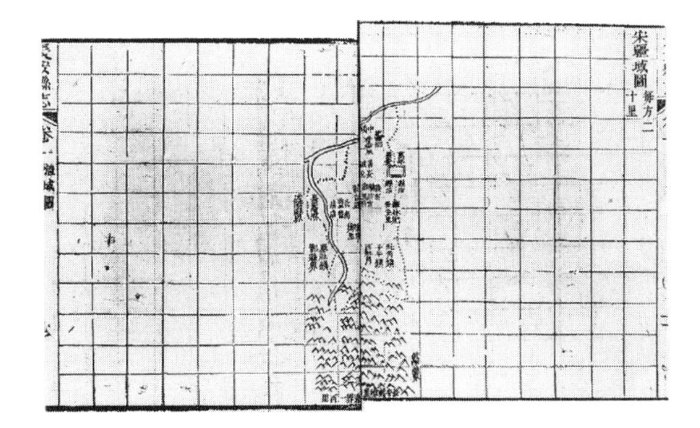

Figure 1.28. *Territories* (in vicinity of Chang'an in Song period). From *Chang'an xian zhi, juan* 1/11b–12a.

Luo Hongxian's (1504–64) *Guang yutu* (Enlarged terrestrial atlas) of circa 1555 is the oldest printed Chinese set of maps on grid surfaces

FIGURE 1.29. Map of Shaanxi province. Luo Hongxian (1504–64), *Guangyu tu, juan* 1/1b-2a, ca. 1555. After Walter Fuchs, *The Mongol Atlas of China,* p. 7. Courtesy of *Monumenta Serica.*

(figure 1.29).[42] The lineage of Luo's atlas is traced back in time to Zhu Siben's (1273–1337) *Yutu* (Map of territories) of 1320, to Song official scholar cartographer and scientist Shen Kuo's (1031–95) work, and ultimately to Pei Xiu, whose *Yu gong* is believed to have contained a map of China on a grid surface.[43]

The use of the grid as the surface on which to begin drawing a map may have a more deep-rooted significance to the Chinese than the obviously potent visual compatibility of regularly shaped enclosed spaces described in the text about the Zhou ruler's city and pictures of it in illustrated classics. A flat surface with a grid is the locus of a Chinese child's initial encounter with his own culture. From the day a Chinese begins to write it is learned that both simple and complex characters must conform to uniformly sized boxes defined by a grid.[44] The practice is contin-

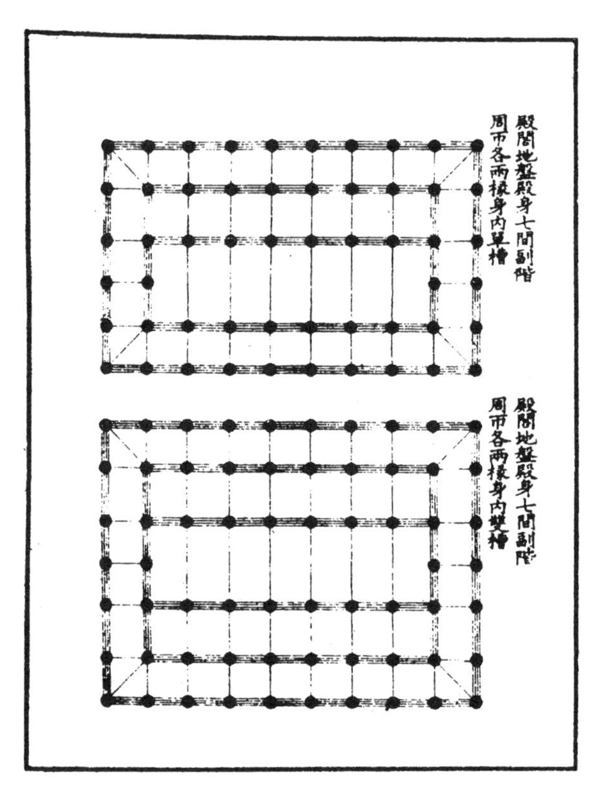

FIGURE 1.30. *Chang'an xian zhi* [Record of Chang'an district], 1815, *juan* 4/7b-8a.

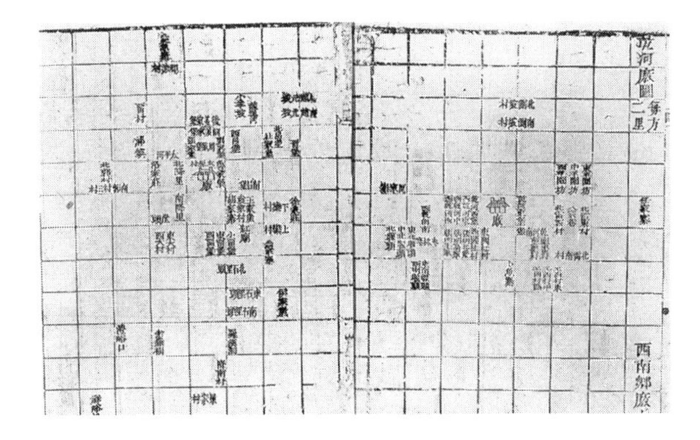

FIGURE 1.31. Column grid plans of timber frame halls. From Li Jie, ed., *Yingzao fashi* [Building standards], 1103, reprinted 1167 (Taibei, 1974), *juan* 31, 7: 30.

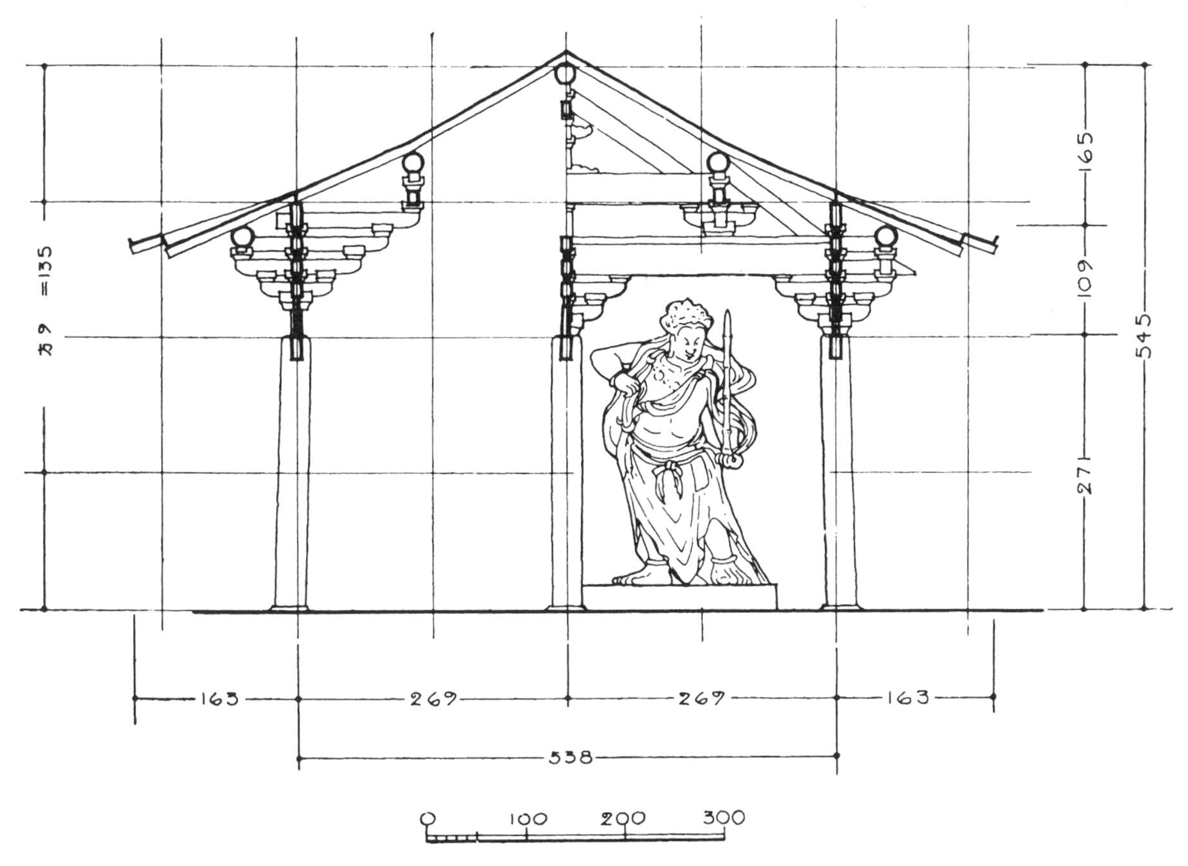

FIGURE 1.32. Sectional drawing of Shanmen [Front Gate], Dule Monastery, 984. From Chen Mingda, "Dulesi Guanyin Ge, Shanmen, jianzhugou tu fenxi" [Pictorial analysis of the timber frames of Guanyin Pavilion and the Front Gate of Dule Monastery], *Wenwu yu kaogu lunji* (Beijing, 1986), p. 346. The unit of measure in this drawing is *fen*. 1 *fen* = ⅓cm.

ued by renowned calligraphers for complicated texts that appear in such places as colophons appended to paintings and funerary inscriptions or historical records carved on stone stelae. Writing and printing of characters occurs on grid surfaces, especially in official publications, even when the placement of characters may, like the plan in figure 1.24, seem otherwise illogical (figure 1.30). Thus again cartography joins the elite scholarly attributes of calligraphy and game playing in the common execution on a surface of parallel and perpendicular lines. Yet cartography and calligraphy share a feature that distinguishes them from painting, poetry writing, and the other scholarly virtues. The former two began as purely utilitarian means of record keeping.[45] In their more practical functions, cartography and calligraphy join building construction.

The backbone of Chinese timber frame construction is also a column grid, whose plan is thus easily plotted on a grid surface (figure 1.31). One can also find examples of the elevation of

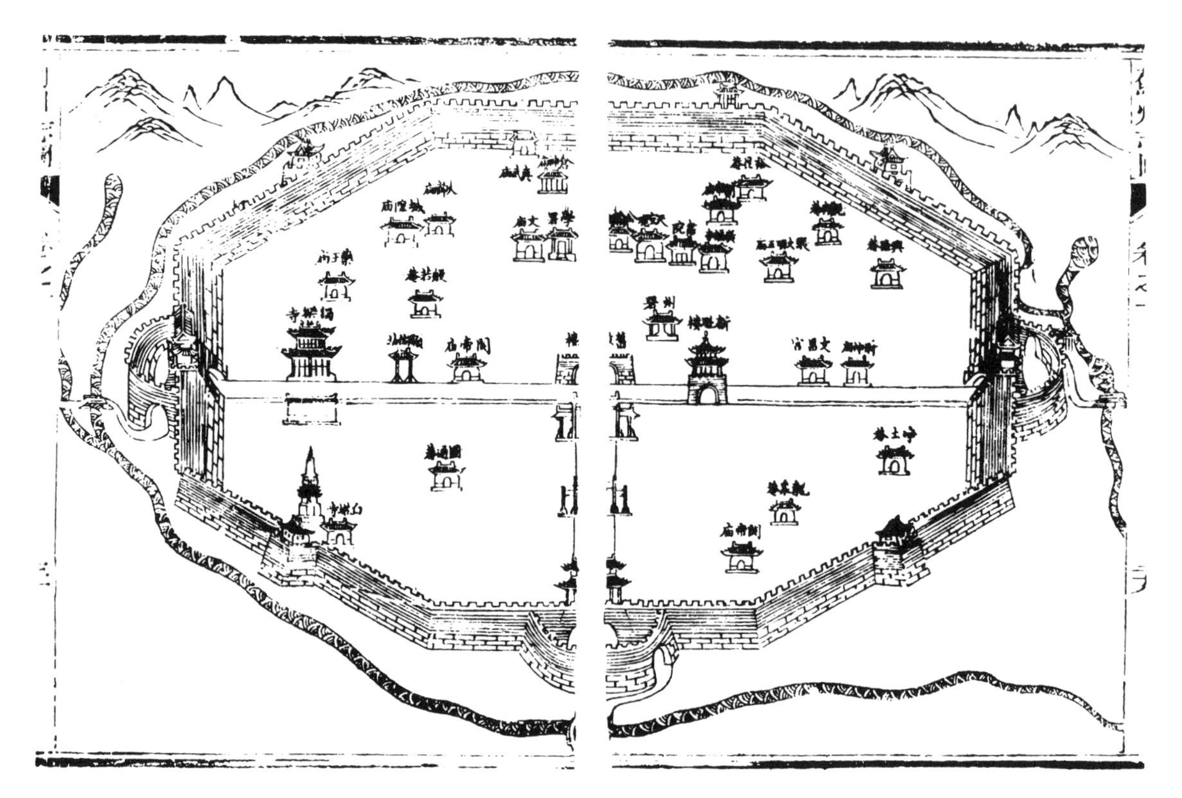

FIGURE 1.33. *Newer Walled City of Jizhou.* From Wu Tang and Zhang Chaozong, *Jizhou zhi* [Record of Ji prefecture], 1704, (Taibei, 1968), 1: 58–59.

the section of a Chinese wooden building drawn on a grid surface (figure 1.32). Unlike calligraphy and cartography, architectural design never became an attribute of China's scholarly elite. One does not find drawings of even the most famous prefectural or district buildings in isolation in local records—buildings are always shown as part of building complexes or cities. The logic of the inclusion of the description of Wangcheng in the Record or Trades (*Kaogong ji*) section of the *Rituals of Zhou* must have been because its primary function was as a utilitarian art.

Still, one can find maps of Chinese cities that do not show orthogonal lines and are not made on grids. These plans most often depict the most important town of a district or prefecture and are preserved in their local records. Yet the exclusion of one set of cartographic conventions gives way to different ones. Maps of towns in these local records are almost invariably walled, mountains are located in the north, cities are surrounded by water, and major north-south and east-west streets cross in town centers (figure 1.33). At the crossroads, one knows from texts, were bell towers, the timekeeping devices crucial to the proper order of the town.[46] A town map that anticipates these rules can be found in relief sculpture from a second-century A.D. tomb (figure

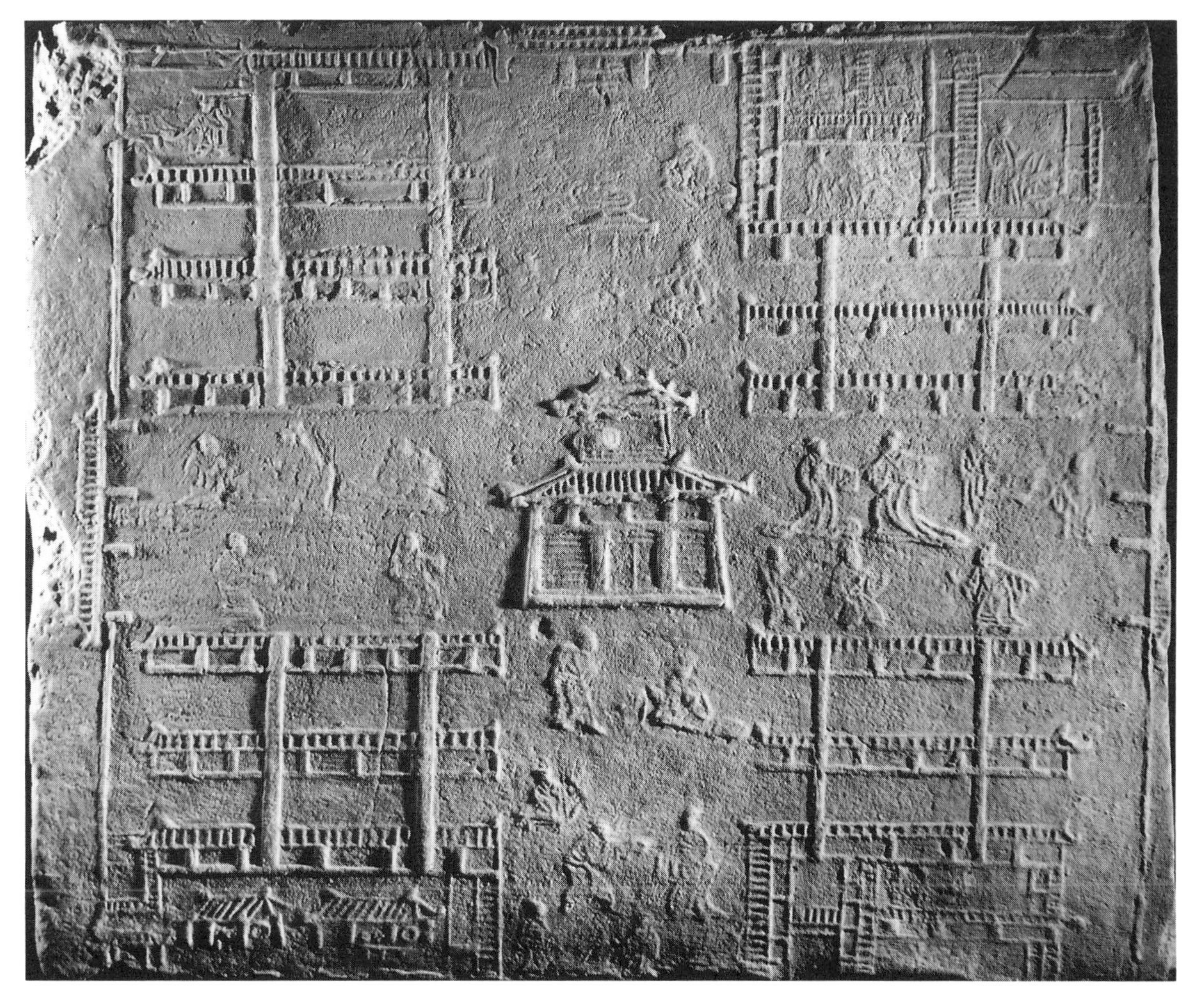

FIGURE 1.34. Relief of City Streets, Chengdu. Tomb tile from Chengdu, Sichuan, Eastern Han. After Lucy Lim, *Stories from China's Past* (San Francisco: The Chinese Culture Foundation of San Francisco, 1987), p. 101.

1.34). No map of a Chinese city could escape correspondence to a set of criteria associated with some textual or symbolic purpose. Mapmaking in premodern China was not a technical exercise striving toward accuracy but an art among elite arts in which service of state and associated lofty purpose of virtue can supersede truth.

Unless it was painted on a tomb or temple wall, a map and its making were the property and activity of the Chinese court. Therefore, like so many other aspects of traditional Chinese culture, the plan of a city was subject to unchangeable rules that could be traced to the classical age and its texts and justifiable—whatever form it took—by service of a higher purpose. Placing cartography of cities within the sphere of traditional values gave it greater status than the perishable material on which it was made (the age of a map does not give it worth—what it depicts gives it pedigree) or the transitory material of the buildings that might be plotted on it. The reality of the map of a Chinese city lay in the regal system it symbolized in which there was always a place for a ruler to sit facing south in the center.

Notes

1. I was informed of this feature of 1960s maps of Taibei by John Wills in 1991 during a helpful conversation about the subject of this paper.

2. See, e.g., Dong Jianhong et al., *Zhongguo chengshi jianshe shi* [History of Chinese city construction] (Beijing, 1982); He Yeju, *"Kaogong ji" yingguo zhidu yanjiu* [Research on the building system via the *Kaogong ji*] (Beijing, 1985); He Yeju, *Zhongguo gudai chengshi guihua shi luncong* [Discussion of the history of ancient Chinese city planning] (Beijing, 1986); Murata Jiro, *Chūgoku no teito* [Chinese imperial cities] (Kyoto, 1981); and Wu Liangyong, "A Brief History of Ancient Chinese City Planning," *Urbs et Regio* 38 (Kassel, 1986). My own research comes to different conclusions. See, e.g., "Why Were Chang'an and Beijing So Different?" *Journal of the Society of Architectural Historians* 45, 4 (1986): 339–57 and *Chinese Imperial City Planning* (Honolulu, 1990). Ideas from these publications are summarized in this section. Illustrations of all city plans discussed but not published here are available in the book.

3. In the period referred to in the *Rituals of Zhou, jiangren* was an official title. One of the *jiangren*'s primary responsibilities was royal building projects, including supervision of artisans. In recent times, *jiangren* has been translated as architect, but there was no premodern Chinese counterpart for an architect.

4. For a French translation of this passage from *juan* 2 of the *Kaogong ji* [Record of trades] section of the *Rituals of Zhou* see Edouard Biot, translator, *Le Tcheou-li*, 2 vols. (Paris, 1851), 2: 553–59. Paul Wheatley's translation in *Pivot of the Four Quarters* (Chicago, 1972), 426, 411, is somewhat different.

5. Its original source is *Henan zhi* [Record of Henan province], Yuan (1279–1368) period, which no longer survives. The earliest reprinting of the record and maps is found in *Yongle dadian* [Encyclopedia of the *Yongle* reign period], *juan* 9561, first illustration, published in 1408.

6. Sometimes there is a third walled area in Chinese capitals. Known as *huangcheng* (imperial or administrative city), it may enclose the palace city or it may be separate but adjacent to it. Government offices are contained within *huangcheng*.

7. On the excavation, see "Luoyang Jianbin Dong Zhou chengzhi fajue baogao" [Excavation report on the Eastern Zhou city at the Jian riverbank, Luoyang], *Kaogu xuebao* no. 2 (1959): 15–36, and "Zhongguo Kexueyuan Kaogu Yanjiusuo yijiuliushi-nian tianye gongzuo de zhuhao shouhuo" [Important gains in fieldwork of the Archaeology Research Institute of the Chinese Academy of Sciences in 1960], *Kaogu* no. 4 (1961): 214–18.

8. This was the subject of my dissertation, "Imperial Architecture under Mongolian Patronage: Khubilai's Imperial City of Daidu" (Harvard, 1981) and "The Plan of Khubilai Khan's Imperial City," *Artibus Asie* 44, 2/3 (1983): 57–73. Addional evidence about the plan of Dadu and the role of Khubilai's close Chinese advisor, Liu Bingzhong, in the choice of this plan, is found in Hok-lam Chan, "A Mongolian Legend of the Building of Peking," *Asia Major* 3, 2 (1990): 63–93.

9. Song Lian (1310–81), ed., *Yuan shi* [History of the Yuan dynasty] (Beijing, 1976), 1879–80.

10. Xu Song (1741–1848), compiler, *Song huiyao jijigao* [Rules and regulations of the Song] (Beijing, 1936), *juan* 188/8a.

11. On Song Kaifeng, see Edward Kracke, "Sung K'aifeng: Pragmatic Metropolis and Formalistic Capital," in *Crisis and Prosperity in Sung China*, ed. John Haeger (Tucson, 1975), 49–77.

12. Arthur C. Moule and Paul Pelliot, *Marco Polo: The Description of the World* (London, 1938), 212.

13. Joseph Rykwert, *The Idea of a Town* (Princeton, N.J., 1976), 26.

14. For a discussion of these aspects of ancient Roman planning and illustrative evidence see ibid., chap. 2.

15. At least, southern orientation is implied in the suggestion that cities (whose sites slope eastward) should face away from the north. See Aristotle *Politics* 7.10. See also Rykwert, *Idea of a Town*, esp. 41–44.

16. On haruspex see L. Bouke van der Meer, "Iecur Placentinum and the Orientation of the Etruscan Haruspex," *Bulletin Antieke Beschaving* 54 (1979): 49–64, and his *The Bronze Liver of Piacenza: Analysis of a Polytheistic Structure* (Amsterdam, 1987). I thank Ann B. Brownlee for these two references. On Chinese scapulimancy and plastromancy, see David Keightley, *Sources of Shang History: The Oracle Bone Inscriptions of Bronze Age China* (Berkeley and Los Angeles, 1978), 1–27.

17. Rykwert, *Idea of a Town,* 48–49, whose source is Pliny, *Natural History* 18.

18. See Rykwert, *Idea of a Town,* 35–36. For the excavation report, see Paulo Claudio Sestieri, "Il Sacello-Heroon Posidonate," *Bolletino d'Arte* 50 (1955): 53–61.

19. The first mention of the nine bronze tripods is in the *Zuo zhuan* (Commentary of Zuo). The vessels have special notoriety because they were nearly lost by a third-century B.C. emperor. That incident is depicted in Han-dynasty relief sculpture from the Wu Family Shrines and Xiaotangshan, both in Shandong province, and elsewhere. On the nine bronze vessels see Wu Hung, *The Wu Liang Shrine* (Stanford, Calif., 1990), 92–96.

20. For a translation of the pertinent passage from the *Zuo zhuan* and more discussion of the nine bronze cauldrons, see Joseph Needham, *Science and Civilization in China* (Cambridge, 1959), 3:503–4. Needham's discussion of Chinese geography and cartography (pp. 497–590) remains an excellent survey of the subject.

21. On the bronze plate, see "Hebeisheng Pingshanxian Zhan Guo shiqi Zhongshanguo muzang fajue jianbao" [Brief report on the digging at the necropolis of the Zhongshan kingdom from the Warring States period in Pingshan, Hebei], *Wenwu* no. 1 (1979): 1–31; Fu Xinian, "Zhan Guo Zhongshan Wang Cuo mu chutu de 'zhaoyu tu' ji qi lingyuan guizhi de yanjiu" [Research on *zhaoyu tu* from the tomb of King Cuo of the Zhongshan kingdom of the Warring States period and the arrangement of surrounding tombs], *Kaogu xuebao* no. 1 (1980): 97–118; and Yang Hongxun, "Zhan Guo Zhongshan Wang ling ji 'zhaoyu tu' yanjiu" [Research on the tomb and *zhaoyu tu* of the Zhongshan kings of the Warring States period], *Kaogu xuebao* no. 1 (1980): 119–38. For a color picture of the bronze plate, see Cao Yanru et al., *Zhongguo gudai ditu ji* [An atlas of ancient maps in China] (Beijing, 1990), 1: pl. 1.

22. See He Shuangquan, "Tianshui Fangmatan Qinjian zongshu" [Survey of the Qin period bamboo slips from Fangmatan in Tianshui], *Wenwu* no. 2 (1989): 23–31. There is more than one idea about the occupant of this tomb. When reports of the excavation first came out, he was believed to be a military official and later diviner with the given name Dan. In 1990 Li Xueqin suggested the character Dan did not refer to his name. See Li Xueqin, "Fangmatan jianzhong de zhiguai gushi" [Strange Stories on Slips from Fangmatan], *Wenwu* no. 4 (1990): 43–47. I thank David Buck for allowing me to read and quote from the unpublished manuscript, David Buck and Hu Bangbo, "Ancient Chinese Maps from Third and Second century B.C. Tombs in Gansu," and for helpful discussions about the Fangmatan maps.

23. See Buck and Hu, "Ancient Chinese Maps," esp. 17–20.

24. English discussions of the Mawangdui maps are found in Annelise G. Bulling, "Ancient Chinese Maps: Two Maps Discovered in a Han Dynasty Tomb from the Second Century B.C.," *Expeditions* 20, 2 (1978): 16–25; Mei-ling Hsu, "The Han Maps and Early Chinese Cartography," *Annals of the Association of American Geographers* 68, 1 (1978): 45–60; Kuei-sheng Chang, "The Han Maps: New Light on Cartography in China," *Imago Mundi* 31 (1979): 9–17; and R. R. C. de Crespigny, "Two Maps from Mawangdui," *Cartography* 11, 4 (1980): 211–22. Chinese bibliography is provided in the notes of these articles. Interestingly, a star chart was also excavated in Mawangdui Tomb 3. Buck and Hu (esp. 27–31) suggest this chart may have been used for prognostication. It may not be coincidence that maps were found together with objects for prognostication in the second-century B.C. tomb. The suggestion is that the same group in society that had access to military maps either were allowed to be in possession of (at least for burial), or were readers of, devices of fortune-telling.

25. For illustrations of the military and topographical maps from Mawangdui, see Cao Yanru et al., *Atlas of Ancient Maps of China,* pls. 20–27.

26. Allyn Rickett, trans., *Guanzi* (Princeton, N.J., 1985), 389.

27. For more on the relation between these and other texts and cartography, see Needham, *Science and Civilization in China,* vol. 3, esp. 500–520.

28. Mei-ling Hsu, "The Han Maps and Early Chinese Cartography," 45, makes the point that Han maps are either of the descriptive or analytical tradition.

29. The source of the story is the problematical text, *Han Wudi gushi* [Stories of emperor Han Wudi]. For a version attributed to Ban Gu (first century A.D.), see *Lidai xiaoshi* (Stories through the generations), vol. 1, *juan* 4/1a–6a. For more on this problematical text, see Thomas E. Smith, "Where Chinese Administrative Practices and Tales of the Strange Converge: The Meaning of *Gushi* in the *Han Wudi Gushi, Early Medieval China* 1 (1994): 1–33. The story is related in Michael Sullivan, *The Birth of Landscape Painting in China* (Berkeley and Los Angeles, 1962), 36.

30. Zhang Yanyuan, *Lidai minghua ji* [Record of painters of successive generations], 847, reprint (Taibei, 1971), *juan* 4, 165–66. See also Sullivan, *Birth of Landscape Painting in China,* 36–37.

31. On Chinese painting of the third through sixth centuries, see, e.g., Sullivan, *Birth of Landscape Painting in China,*

esp. chap. 2, and Sickman and Soper, *The Art and Architecture of China* (Harmondsworth, 1971), 129–34.

32. Translations and some discussion of Pei Xiu's Six Principles appear in Edouard Chavannes, "Les deux plus anciens spécimens de la cartographie chinoise," *Bulletin de l'École Française d'Extrême-Orient* 3 (1903): 214–47, and esp. 241–43; Needham, *Science and Civilization in China,* 3:539–40, from which the translations here are adapted; Sullivan, *Birth of Landscape Painting in China,* 36; Cheng-siang Chen, "The Historical Development of Cartography in China," *Progress in Human Geography* 2, 1 (1978): 101–20, and esp. 103; and Cao Yanru et al., *Atlas of Ancient Chinese Maps,* 3.

33. On the Six Laws of Xie He see, e.g., Arthur Waley, *An Introduction to the Study of Chinese Painting* (New York, 1923), 72; Osvald Siren, *The Chinese on the Art of Painting,* (Beijing, 1936), 19, 219; Shio Sakanishi, *The Spirit of the Brush* (London, 1939), 43–45; Alexander Soper, "The First Two Laws of Hsieh Ho," *Far Eastern Quarterly* 8 (1949): 412–23; W. R. B. Acker, *Some T'ang and Pre-T'ang Texts on Chinese Painting* (Leiden, 1954), 3–32; Sullivan, *Birth of Landscape Painting in China,* 105–10; James Cahill, "The Six Laws and How to Read Them," *Ars Orientalis* 4 (1961): 372–81; Wen Fong, "The First Principles of Hsieh Ho," *National Palace Museum Quarterly* 1, 3 (1967): 7–18; Sickman and Soper, *Art and Architecture of China* 133; and John Hay, "Values and History in Chinese Painting, I: Hsieh Ho Revisited," *RES* 6 (1983): 73–111.

34. On the Han tomb and pictures of its wall paintings, see four articles in *Wenwu* no. 1 (1974); Gai Shanlin, *Helinge'er Hanmu bihua* [Wall paintings from the Han tomb at Helinge'er] (Hohhot, 1978), and A. Bulling, "An Eastern Han Tomb at Holin-ko'ehr," *Archives of Asian Art* 31 (1977–78): 79–103.

35. On these tombs and their wall paintings, see Mary H. Fong, "Four Chinese Royal Tombs of the Early Eighth Century," *Artibus Asiae* 35, 4 (1973): 307–34 and "T'ang Tomb Wall Paintings of the Early Eighth Century," *Oriental Art* 24, 2 (1978): 185–94; and Jan Fontein and Tung Wu, *Mural Painting Han to Tang* (Boston, 1976). Chinese bibliography is provided in notes of these works.

36. See, e.g., Fu Xinian, "Tang Chang'an Daminggong Hanyuandian yuanzhuang de tantao" [Research on the original appearance of the Hanyuan Hall of Daming palace complex at Tang Chang'an], *Wenwu* no. 7 (1973): 30–48.

37. Fu Xinian, "Tangdai suidao xingmu de xingzhi gouzao he suofanying de dishang gongshi" [Above-ground palace architecture as reflected on the walls of Tang tomb paths], *Wenwu yu kaogu lunji* (Beijing, 1986), 322–43.

38. Fu Xinian, "Shanxisheng Fanshixian Yanshansi nandian Jindai bihuazhong suohui jianzhu de chubu fenxi" [Early stages of analysis of Jin-period wall paintings of architecture in the south hall of Yanshan Monastery, Fanshi county, Shanxi], *Jianzhu lishi yanjiu* (1982): 119–51. The source of the description is Fan Chengda (1126–93), *Lanpei lu* [Record of grasping the reigns], available in various reprinted versions including *Zhibuzuzhai congshu,* vol. 215 (Taibei, 1966).

39. This painting has been studied by Hibino Jōbu, "Tonkō no Godai-sanzu ni tsuite" [On a map of Mt. Wutai at Dunhuang], *Bukkyō bijitsu* 34 (1958): 75–86; Ernesta Marchand, "The Panorama of Wu-t'ai Shan as an Example of Tenth Century Cartography," *Oriental Art* 22, 2 (1976): 158–73; and Dorothy Wong, "A Reassessment of the *Representation of Mt. Wutai* from Dunhuang Cave 61," *Archives of Asian Art* 46 (1993): 27–52.

40. Verbal directions from the local population were the standard means of obtaining travel information by pilgrims in medieval East Asia. Illustrated biographies and legends of monks contain paintings of this subject. See e.g., scroll 3 of the *Shigi-san engi* [Legends of Mt. Shigi], twelfth century, in the Chōgosonshi-ji, Nara, which shows the journey of a nun in search of her brother, the monk Myōren. In my travels through Shanxi and Hebei in 1986 and 1987 I still found the directions of locals more valuable than available city or provincial maps.

41. On Han Chang'an, pre- and postexcavation maps of it, and discussion of the city in texts and secondary sources, see Steinhardt, *Chinese Imperial City Planning,* 54–68.

42. Stelae on which maps are carved on grid surfaces survive from the Song dynasty (960–1279). Three years after this paper was written, J. B. Harley and D. Woodward, *The History of Cartography,* vol. 2, bk. 2, *Cartography in the Traditional East and Southeast Asian Societies* (Chicago, 1994) was published. Many of the maps discussed here are also discussed in that book. The authors of the chapters on Chinese cartography do not give Pei Xiu as much credit for the concept of a grid surface as I do. I plan to discuss this topic in a future publication.

43. On Zhu Siben's map, see Walter Fuchs, *The Mongol Atlas of China* (Beijing, 1932). For more on Zhu Siben, Shen Kuo, and other key figures in the history of Chinese cartography, see the section on cartography in Needham, *Science and Civilization in China,* vol. 3, cited above; Harley and Woodward, eds., *The History of Cartography,* vol. 2, bk. 2; and Wang Yong, *Zhongguo ditu shigang* [History of Chinese maps] (Beijing, 1958).

44. On the practical principles of Chinese calligraphy, see

the introduction to Tseng Yu-ho Ecke, *Chinese Calligraphy* (Philadelphia, 1971). Examples of how characters are formed in evenly sized boxes are shown on the first page of the introduction.

45. Aspects of the relationship between painting and calligraphy are explained in Lothar Ledderöse, "Subject Matter in Early Chinese Painting Criticism," *Oriental Art* 19, 1 (1973): 69–83.

46. Tanaka Tan, "Jujiro ni tatsu hoji rokaku" [Towers for keeping time at the crossroads], *Chyamus* 5 (1983): 20–22.

TWO

Mapping the City: Ptolemy's *Geography* in the Renaissance

Naomi Miller

City Maps

One of the earliest known collections of city maps suddenly appears in three manuscripts of Ptolemy's *Geography*. Produced in Florence in the second half of the fifteenth century, these city maps were appended to the map of the world and the regional maps. The group is unique in that it includes both Italian and Muslim cities—namely, Milan, Venice, Florence, Rome, Constantinople, Damascus, Jerusalem, Alexandria, and Cairo. All three codices are based on Jacopo d'Angelo's translation of the *Geography* from Greek to Latin, completed in 1406.[1] In addition, they share the same scribe, the French copyist Ugo Comelli da Mezières, and the same artist, Pietro del Massaio.[2]

It was Massaio (active 1458–72) who first added these city maps to the manuscripts of the *Geography* under discussion. In comparison with the standard Ptolemaic maps, those of Massaio are distinguished by improved orientation and detail and appear to be based on more precise instrumental observation and geographical knowledge. Like most of the early mapmakers, Massaio was trained as a painter-illuminator, probably skilled in mathematical cartography and conversant with Ptolemaic systems of projection. He presumably worked from old models and under the supervision of humanist advisers and patrons, for there is no evidence that he designed these maps.[3] In their execution the maps reflect pictorial elements that emerged from the new art of perspective. Emphasis on measurement and proportion appears to have resulted in clearer depictions of individual buildings and more accurate relationships between buildings and their surroundings.

Two of the manuscripts are now in the Vatican Library and are securely dated, namely *Vat. lat. 5699*, 1469, commissioned by humanist Niccolo Perotti, archbishop of Siponto, and *Vat. Urb. lat. 277*, 1472, made for Federigo da Montefeltro of Urbino. The date of the third manuscript, *Paris B.N. Lat. 4802*, made for King Alphonso of Naples, is more problematic. Some of the city maps in it contain structures only built toward the end of the century, putting in question the proposed date of circa 1456, and/or reinforcing the possibility of continuous updating.[4] In addition to the cities cited above, a map of Adrianopolis (the modern Edirne), capital of Byzantium until the Turkish conquest, is included in the Paris codex, whereas a map of Volterra is appended to the Urbino manuscript.

The focus of this study is on the city maps in the Urbino codex, viewed singly and collectively. But the very presence of these maps in the *Geography* compels us first to seek a justification for their inclusion. It is my principal hypothesis that the city maps are primarily a manifestation of an expanding worldview, as exemplified by the proliferation of Ptolemaic manuscripts in the Renaissance. More specifically, the incorporation of these maps in the *Geography* acknowledges an updating of the only extant fifteenth-century atlas and, by means of the information conveyed, serves the growing demands for trade and travel in an age of conquest and exploration. Although interest in the Ptolemaic maps was generated initially by the rediscovery of ancient texts accompanied by scholarly discourse, the informed citizen could also appreciate the connection of these maps with charts created for military purposes and as guides for religious pilgrimages. Here was the possibility of broadening one's vistas beyond the confines of the city on the Arno, where these maps were created.

Ptolemy's *Geography*

A rationale for the introduction of city maps, namely, to bring the manuscripts up-to-date, may be provided by Ptolemy himself. The famed astronomer, mathematician, geographer, and man of learning lived in Alexandria circa A.D. 100–170, when the Roman empire was at its height. Whereas his masterpiece, the *Almagest,* charts the celestial sphere, his *Geography* (ca. A.D. 150), an undertaking to map the entire known world, may be deemed its terrestrial counterpart. To this end, Ptolemy created the first atlas, composed of a gazetteer and a collection of maps. At the beginning of book 1, Ptolemy sets forth his goals, the first being to give an adequate representation of the known world. As part of his strategy, he immediately emphasizes the difference between geography—the representation of the *intera ecumene* (entire world) complete with regions and their general features—and chorography—the representation of small parts of this world. Geography deals with "the larger towns and the great cities, the mountain ranges and the principal rivers," whereas chorography treats "the smallest conceivable localities such as harbors, farms, villages, river courses and such like." It is not unimportant, especially given the reception of Ptolemy in the Renaissance, that his text then compares the process of making a map to that of painting a picture: "For as in an entire painting we must first put in the larger features, and afterward those detailed features which portraits and pictures may require, giving them proportion in relation to one another." This analogy with pictorial representation is stressed in the function given to chorography: "to paint a true likeness and not merely to give exact positions and size."[5] Supplementing the knowledge of mathematics, true proportion, and scale necessary for geography is an emphasis on artistic skill as manifest in early Roman landscape painting. May we not posit that Ptolemy's ideas of painting were born in the poetic, if unscientific, views of Roman frescoes that adorned the walls of a first-century A.D. Roman villa with the adventures of Ulysses, or in analogous works in the villas of Pompeii, Boscoreale, and Herculaneum?

Vat. Urb. lat. 277: Form and Content

Urb. lat. 277 is among the eighteen-hundred-odd manuscripts from the library of the duke of Urbino that in 1658 became a part of the Vatican collections under Pope Alexander VII. Formed largely by Federigo da Montefeltro between 1468 and his death in 1482, it was the humanist library par excellence. Amassing fame and fortune as governor and knight *condottiere,* Federigo was determined to build the finest Renaissance palace and to equip it with the most comprehensive library.

Even among the plethora of magnificent volumes in the duke's collection, this Ptolemy atlas is outstanding. Produced in Vespasiano da Bisticci's workshop, the main calligraphic bookshop and supplier of manuscripts in the second half of the fifteenth century, it is contemporary with the first printed editions of Ptolemy then in preparation. Surely this Latin Ptolemy, one of some seventy manuscripts made expressly for Federigo, must have occupied a particularly important status within the contents of his library, a library that contained one of the oldest surviving Greek manuscripts of the *Geography, Urb. gr. 82.*[6] Dated in the late twelfth or early thirteenth century, this codex made in Constantinople was brought to Florence by Emanuel Chrysoloras in 1397. It became the source of the Latin translation by Jacopo d'Angelo here presented, and must have served as the model of *Urb. lat. 277.*

The center of the latter's title page is occupied by the text placard in gold letters proclaiming: IN HOC ORNATISSIMO CODICE CONTINENTUR . COSMOGROPHIE [*sic*] . PTOLEMAEI . VIRI . ALEXANDRINI . DE SITU . ORBIS LIBRI . VIII. EX GRAECO IN LATINUM PER JACOBUM ANGELUM FLORENTINUM . TRADUCTI. (In this most ornate codex are contained the eight books of the Cosmography of Ptolemy of Alexandria on the Geography of the world. Translated from Greek to Latin by Jacopus Angelus of Florence.)

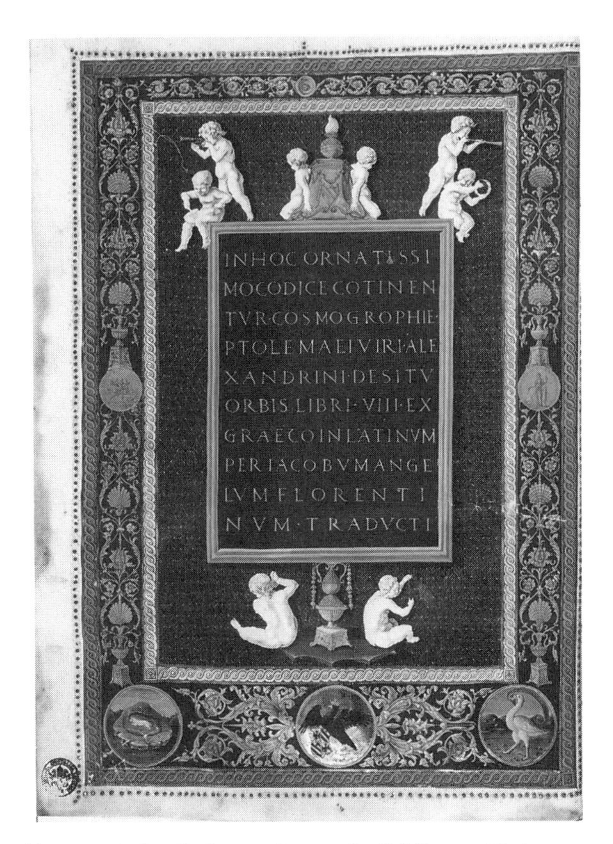

FIGURE 2.1. Ptolemy, *Geography,* Biblioteca Vaticana, urb. lat. 277, 1472, fol. 1. Frontispiece, Francesco Rosselli (photo: Biblioteca Apostolica Vaticana)

Ptolemy's atlas as transmitted in our codex is composed of twenty-seven maps, namely, a map of the world and twenty-six regional maps—ten of Europe, four of Africa, and twelve of Asia—made in accord with his tables of distances (latitudes and longitudes).[7]

Following the eight books, we turn to a bifolium of the world map, identified above: TOTIUS . PARTIS . HABITAE . COGNITAE . QUE . TERRAE . DESCRIPTIO. (Description of that entire part of the world which is inhabited and known.) With its trapezoidal shape, the Ptolemaic map gives rise to a new view of the earth bearing a network of coordinates. Its very shape is a radical departure from the more typical medieval world maps, usually drawn in accord with the scriptures and depicting the world as a disc with Jerusalem at its center. Not as finely rendered as fifteenth-century portolan charts with their intricately drawn harbors and islands, the maps in Ptolemy's *Geography* still incorporate some of the practical data, long used by navigators. Based on that of *Urb. gr. 82,* the world map is also the model for those of the early printed editions.

Seven modern maps, all on bifolia, are appended to the regional maps. Their titles are: DESCRIPTIO . HISPANIE . NOVA; there are similar rubrics for Gallie, Italie, Etrurie, Peloponesi, Crete, and Egypti. According to Almagià, these maps were

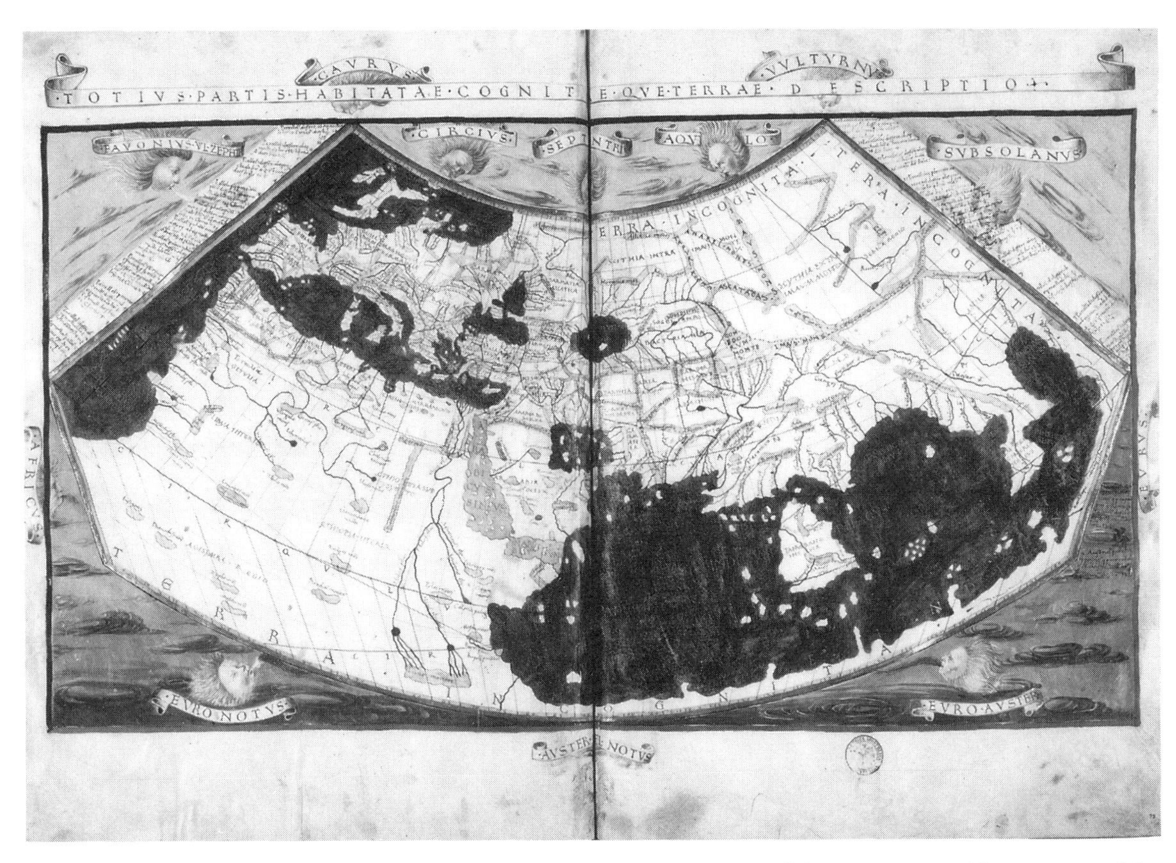

FIGURE 2.2. Ptolemy, *Geography*, Biblioteca Vaticana, urb. lat. 277, 1472, fols. 72v–73r. World Map, Piero del Massaio (photo: Biblioteca Apostolica Vaticana)

probably executed by a group of geographers in Florence and constitute the basis for a modern atlas.[8] He further relates some maps to nautical maps—for example, that of the Peloponnesus is enlarged from a portolan map of the Aegean, with much concern for local color and little attention to verisimilitude.[9]

City Maps: General Characteristics

Surely, the city maps that follow may be related to the same mapping impulse, the same quest for charting the unknown, the same instinct to update the old maps with newly acquired data that prompted the inclusion of the *Tabulae novae* or the seven modern maps in these codices. Even if the city maps had not been made expressly for the Ptolemaic manuscripts, they do represent another cartographic aspect of geography.[10] Herewith we see the chorographic scheme which deals with regional topography and is, according to Ptolemy, to the geographic as the facial features are to the human head.

If we have provided reasons for the inclusion of the city maps in the Ptolemaic codices, other questions remain. How are the city maps drawn? What are their connections with contemporary maps, and with earlier maps? Why are half of the cities here represented Near Eastern metropolises?[11] Why do the means of depicting these cities change from symbolic to "real," from mythical to empirical? How do these maps relate to changing spatial concepts in the Renaissance, artistically as well as politically?

Implicit in these queries are others evoked in studying any maps—questions indigenous to the world of commerce and travel, to military strategy, to political, social, and economic matters, and, above all, to plotting diverse courses of action. What information is given and what omitted? What is the basis for selection?[12] How are the buildings delineated and the open spaces described? Are there any anomalies or surprises? What is stressed and what simplified? What is the purpose of such maps? What images do they convey of the religious order, of cultural and intellectual values, and, more, how do they enlarge our knowledge and understanding of the cities herein depicted, and, ultimately, of the Renaissance worldview?

Prior to examining each map, we will attempt to characterize the group as a whole. To a degree, these maps are in accord with Ptolemy's dictum in book 8, which recommends separate maps of the most populated regions, attention to correct relationships between distance and direction, and selective presentation of data. Certain commonalities are apparent in the draftsmanship of the city maps. All are bird's-eye views rendered in water color and ink on fine vellum, drawn in roughly orthogonal projection, mindful of Brunelleschi's perspectival experiments. Each city is related to its immediate region: topographical features such as waterways, hills, and terrain are emphasized. Large bodies of water, seas, and rivers are painted in deep violet (sometimes opaque, sometimes transparent), and hills and mountains are rendered in sepia-colored wash. Although sometimes implied, there is a virtual absence of a street and road network. Concerning architectural elements, the walls, towers, and gates of the medieval city dominate, the latter even acting as clues to orientation and the location of landmarks. The buildings depicted are usually those that display religious or civic authority and are drawn in approximately the same scale. Some hierarchical considerations, however, may be noted—achieved primarily by the position of the building, but also graphically, in the blue and gilded steeples and domes. Landmarks may have been drawn originally in situ, while ordinary structures are rendered in a conventional manner. To increase the illusion of depth, one side of the building is enhanced by light washes and, further, by hatched lines. Roofs and walls are often painted in rose, pale violet, or blue, and domes in blue with golden spheres. Churches and towers are crowned by glistening globes or pyramidal forms. Open spaces are only vaguely indicated, with a few trees conventionally rendered. We thus glimpse a prospect of the city with its principal monuments, framed, like a picture, by a gilded border. In fact, the title and border are the only vestiges that remain of the highly ornamental earlier maps.

In the Ptolemaic maps, each city (with the exception of Volterra) is identified by its Latin name, written above the map in gold majuscules on violet-tinted banners, in themselves festive icons of honor and glory. The spelling of the city and place names is capricious and varies in the three codices. Latin and Italian are often intermingled, and there is a lack of consistency in church denominations and in the format of abbreviations. Place names are usually inscribed beneath the monuments and above the gates in sepia ink.

Observing the Ptolemaic maps at close range, the main distinctions seem to be between the

Italian cities and those of the Near East. But even among the former, variations are noted. In the maps of Milan and Venice, reliance on older models is evident, whereas in the maps of Florence and, to a lesser degree, that of Rome, it appears as if the drawing of each building is plotted, perhaps on a coordinate system. And, despite the absence of a street network, we can almost sense a connective tissue as we wander from landmark to landmark. In the eastern maps, we approach the city from a bird's-eye view and confront a more coherent panorama, albeit a more abstract prospect, seemingly due to a lack of data. Both types share the same orientation, with south on top, exceptions being Jerusalem, Milan, and Constantinople with west, northwest, and north on top respectively.[13]

The bird's-eye view depicts single monuments obliquely from the north, drawn in empirical two-point perspective. Drawings of different cities are in different modes. Based on actual observation in addition to available prototypes, maps of the Italian cities are far more detailed, comprising a compendium of the most important landmarks, rather than comprehensive views of the whole. As such, they have been referred to as encomiastic views, still within the cartographic spirit of the Middle Ages.[14] On the other hand, aside from Constantinople and Jerusalem, where older models are known, the eastern cities are treated in a relatively summary manner, with far fewer buildings and identifying rubrics. Apparently, the artist felt freer to experiment with new methods of representation, especially in the cities least familiar to him. This is evident in the maps of Damascus, Alexandria, and Cairo, where, despite a degree of accuracy in the general topographic outlines and boundaries, rendering of specific sites and buildings is limited. Some places depicted in these eastern cities duplicate features shown on the maps of Italian cities. Consequently, a different type of map emerges. No longer the medieval collection of monuments, it is not yet a modern map. Buildings are still rendered as insulae, albeit drawn in more correct oblique perspective; and the semblance of a panorama beyond the walls is even reminiscent of townscapes in fourteenth- and fifteenth-century paintings and frescoes.[15] As if to compensate for a lack of data, our description of these maps relies more on extraneous sources, such as pilgrimage accounts and crusader lore. In fact, the very presence of the eastern cities and the monuments therein depicted serves to ally these maps with the journey to the Holy Land.

City Maps: Context

In order to situate the Ptolemaic maps within the context of Renaissance cartography, we will consider developments in Italian and European mapmaking preceding their production. From the medieval legacy of pilgrimage maps, maritime charts, and ideograms, we proceed to maps demonstrating more coherent spatial relationships, culminating in Alberti's lost map of Rome. But, to begin, it is in territorial maps of Padua that we find the strongest affinities to the city views in the *Geography.*

Territorial and Scale Maps

Because the city maps in the Ptolemaic manuscripts approach the end of an era in mapmaking, it is quite startling to find an analogue in a map of Padua inserted in Angelo Portenari, *Della felicità di Padova,* 1623, made by Vincenzo Dotto, a mapmaker skilled in cosmography and architecture.[16] Drawn in pen on paper, the maps depict "Padua surrounded by old walls" and "Padua surrounded by new walls." Executed in the mode of Massaio, emphasizing the circuit of walls and towers and the extensive major and minor systems of canals, the former map is dis-

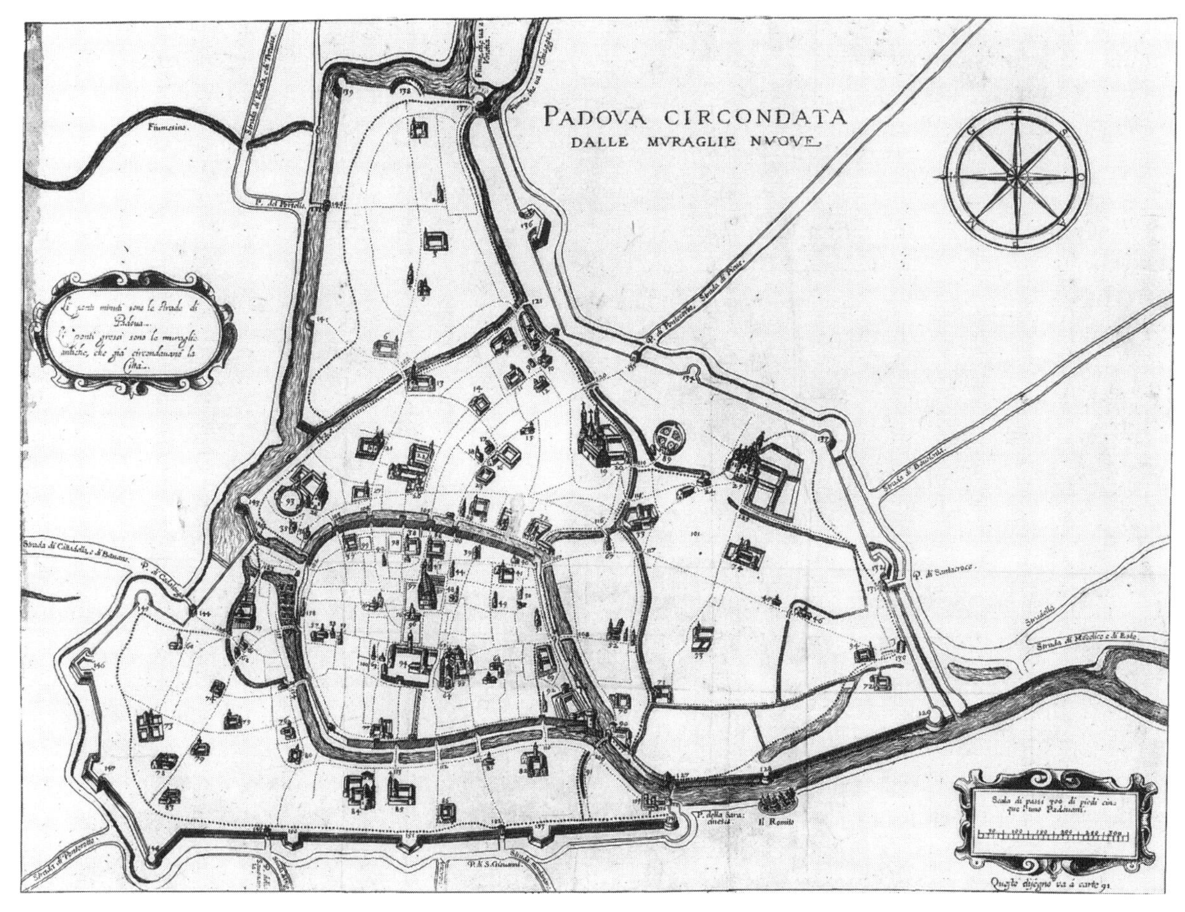

FIGURE 2.3. Vincenzo Dotto, Map, "Padua Surrounded by Old Walls." From A. Portenari, *Della felicitá di Padova*, 1623 (photo: Padua, Museo Civico)

tinguished by its almost exclusive focus on churches and monasteries outside the walls. Buildings are all rendered in axonometrical projection, and street lines are indicated.

Considering the advances in the representation of cities, Dotto's drawing is surprisingly *retardataire*. In this particular context, we may well inquire if his maps were based on a model that preceded or followed the Ptolemaic maps. Evidence for the former is to be found in the territorial maps of Padua that depict the city at the center of a larger district. A perspective view by Annibale Maggi is dated 1449 and is known by a sixteenth-century copy in Milan, Biblioteca Ambrosiana.[17]

The very format of the Maggi map showing the walled city at the center of its territory within a circular orb replicates the medieval map of the world. Padua with its double set of walls is connected to the cities of the Veneto by an elaborate network of inland waterways. Because of its relatively abstract format, the map probably had a celebratory rather than a documentary function, and belongs to that genre related to the praise

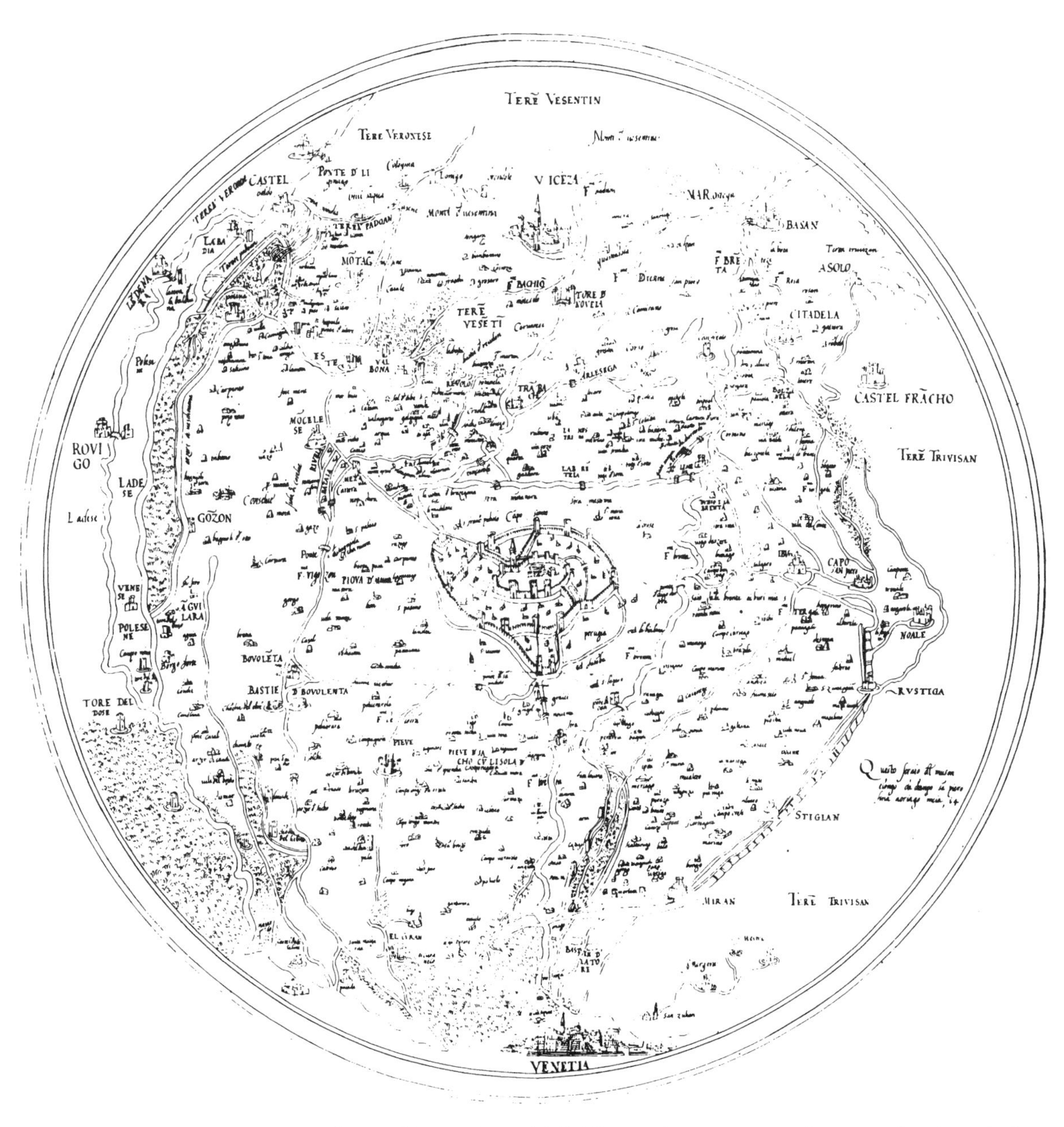

FIGURE 2.4. Annibale Maggi, Padua and its territory, 1449 (Sixteenth-century copy in pen and water color) (photo: Milan, Biblioteca Ambrosiana)

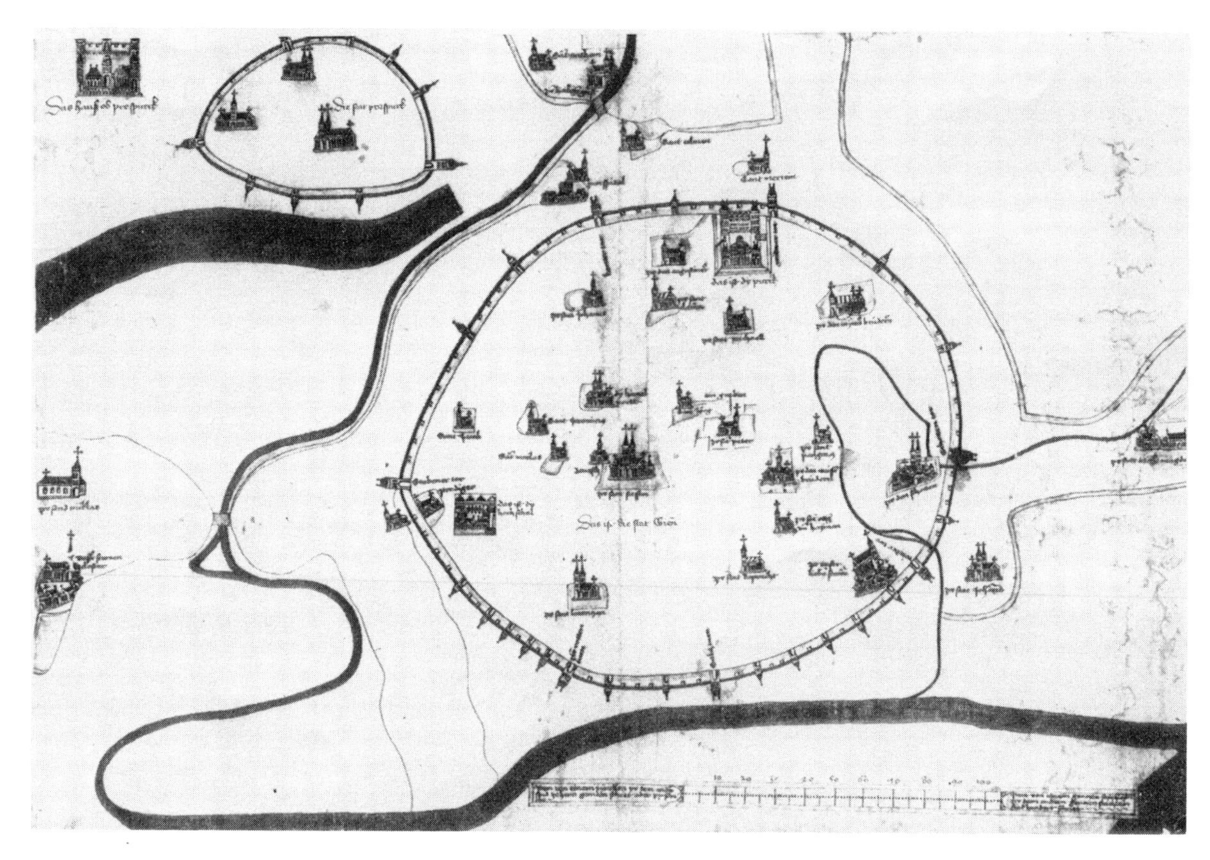

FIGURE 2.5. Vienna, Map, ca. 1422 (mid-fifteenth-century copy), I. N. 31.018 (photo: Vienna Historisches Museum)

of cities.[18] Thus, Padua assumes the spiritual, moral, and civic hegemony in reference to its territory, a sentiment found in a contemporary hagiographic text, the *Libellus* of Michele Savonarola, 1444–47.[19]

Probably based on a lost archetype, continuously updated, Maggi's map had a sufficient level of accuracy to serve as a basis for a map commissioned to Francesco Squarcione by decree of the Council of Ten. Dated 1465, the painter has updated this map on vellum by adding canals in the Paduan basin and new fortifications.[20]

Similar to these maps of Padua is one of Vienna. This later copy of an original, dated 1421–22, represents the earliest known map of that city. Like the Ptolemaic maps, it shows a walled town, drawn in bird's-eye view perspective, here dominated by the Danube and its tributaries. Buildings depicted stress the main churches, set within the plan as isolated entities, since a street

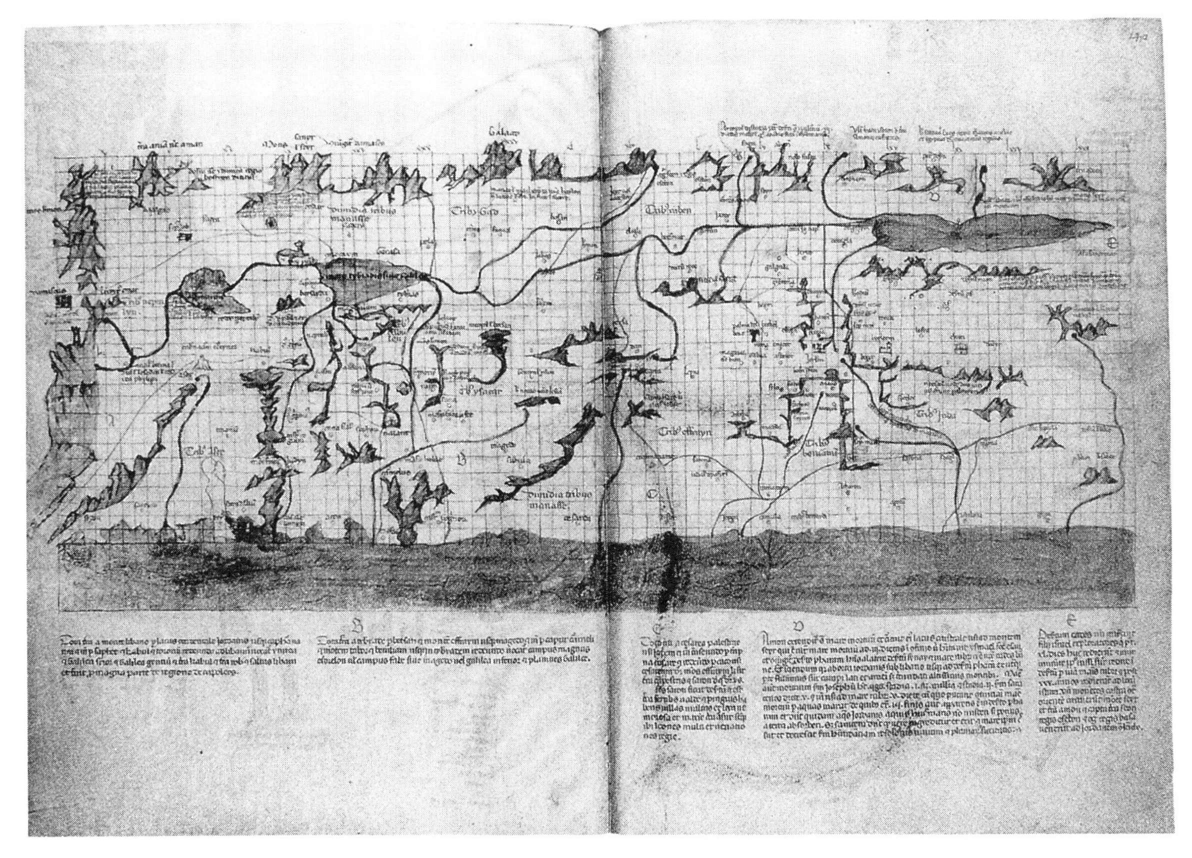

FIGURE 2.6. Pietro Vesconte [attributed], Map of Palestine, from Marino Sanudo, *Liber Secretorum Fidelium Crucis. . .*, ca. 1320, MS. Vat. Reg. Lat. 548, fol. 141v–142v. (photo: Biblioteca Apostolica Vaticana)

network is absent. By far its most extraordinary feature is the scale-bar.[21]

Our search for comparable cultural images has revealed a few maps that seem to stand on the threshold of new cartographic developments. To a degree, their spatial concepts may have been incorporated by the depiction of cities in the Ptolemaic codices. Because the idea of a map drawn to scale is so rare, one cannot but wonder if the fourteenth-century map of Palestine attributed to Pietro Vesconte in Marino Sanudo's *Liber Secretorum Fidelium Crucis* was known to Massaio.[22] Here the Holy Land is drawn on a grid divided into squares, defining points of towns and the measured itineraries between them. One recalls Manetti's posthumous account of Brunelleschi's drawings of ancient monuments using a system of coordinates.[23]

Medieval Precedents

Reviving Ptolemy's *Geography* paved the way for the emergence of modern cartography in a society still dominated by medieval thought. Witness earlier maps, which, to a degree, are based on an

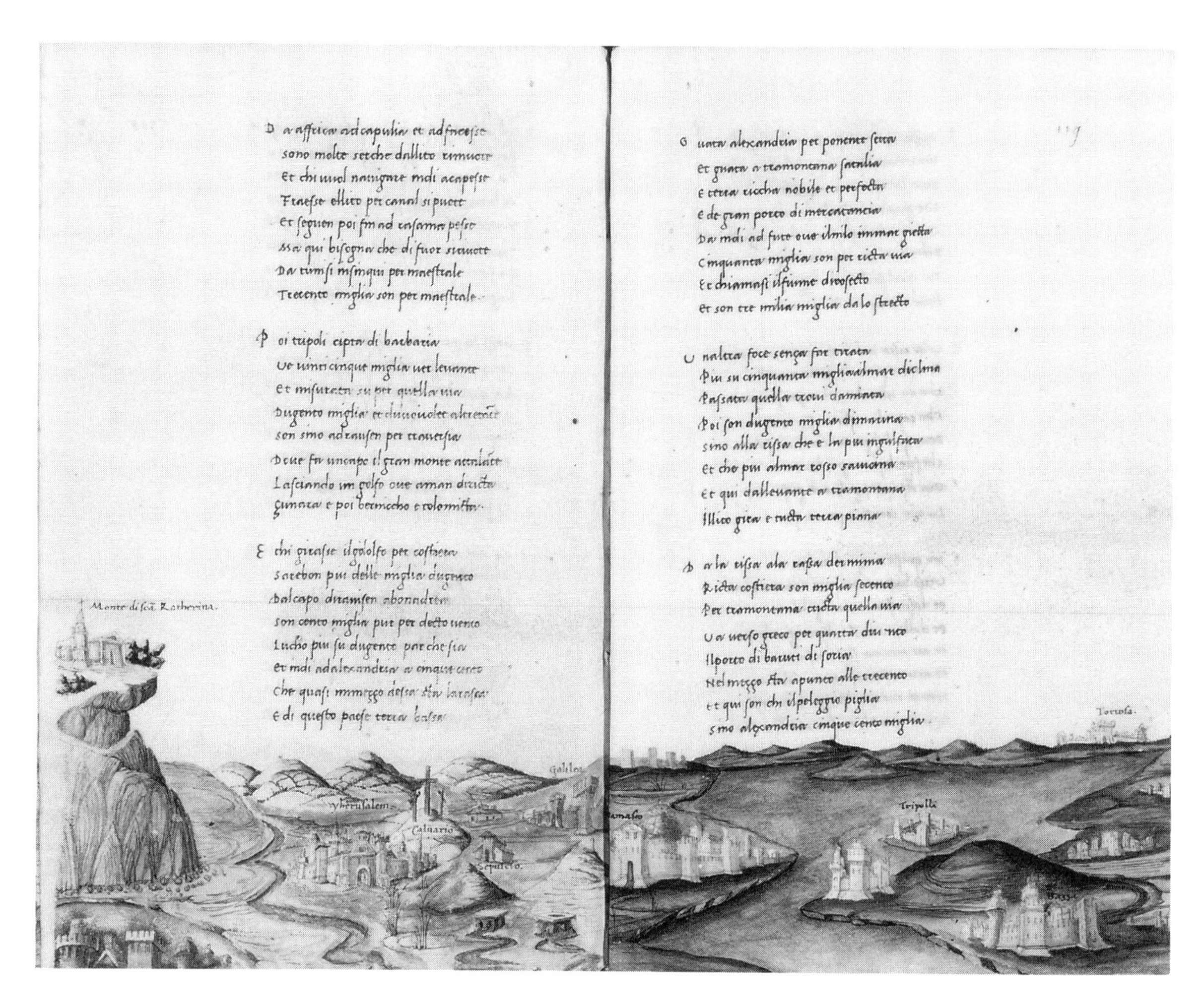

FIGURE 2.7. Goro Dati, *La sphera* (detail), Biblioteca Vaticana, urb. lat. 1754, ca. 1435, fols. 118v–119r (photo: Biblioteca Apostolica Vaticana)

affective experience of the world, a world with Jerusalem at its center, where rubrics are animated by illustrations of biblical events. Coexisting with the renewed interest in the *Geography,* we find both maps that convey a fanciful global concept and portolan charts, carefully describing ports of call and islands, that provide practical information for navigators. And, like the *mappamundi,* which often reflected world history rendered in accord with the scriptures, town views of the Middle Ages presented a conventional grouping of turreted buildings punctuated by campaniles or domes and surrounded by walls, analogous to the walled citadels known from ancient texts. An enclosed mass with little concern for spatial relationships, the city was presented as a totality, its focus was a collection of monuments, its purpose often an exposition of moral exempla.[24] The didactic mode had little need for an accurate presentation of physical topography.

To visualize the older tradition, we may turn to the cosmological poem, *La sphera,* written before 1435 and attributed to the Florentine merchant Goro Dati. Marginal illustrations of the text follow the coastlines in the sailing directions of portolan charts, the shores punctuated by vignettes of cities. Compass bearings indicate harbors, capes, and inlets.[25] Rubrics are within the landward boundaries, whereas the sea remains clear for the rhumb lines that radiate from the compass rose. Few versions of *La sphera* surpass in sheer beauty that of *Vat. Urb. lat. 1754,* a small but precious tome, whose eastern cities include those in our Ptolemaic manuscripts, namely, Cairo, Alexandria, Jerusalem, and Damascus.[26]

Thus, at the dawn of the Renaissance, mapping was strongly tied to medieval cosmological schema, portolan charts, and pilgrimage routes to the Holy Land. Most maps could be classified as city ideograms with a circle of turreted walls familiar from backgrounds of medieval paintings, encomiastic views, or guides providing easy access to the principal monuments of the city. Whereas roads and streets seldom appear, the main urban determinants of a medieval town—walls, towers, waterways—are ubiquitous. Diagrams, building plans, hydrographic projects, boundary images, military and district surveys featuring fortifications and roads, as well as painted representations of landscapes and architecture may have served as sources of inspiration for city maps.[27]

Known precedents for most of the cities included in the Ptolemaic maps are few. As expected, those for the two ideal pilgrimage centers, Jerusalem, the Heavenly City, and Rome, the Eternal City, abound. In fact, the very concept of "city" in the Middle Ages is frequently synonymous with Jerusalem. On medieval maps of the world, it is indicated as the center of the world: "This is Jerusalem. I have set it in the midst of the nations and countries that are round about her (Ezekiel 5:5). Among the cities on these maps, Milan, Venice, and Constantinople have been deemed "new Jerusalems" at times in their histories. Representing the city as a circle, the oft-repeated image of crusader maps conveys a symbolic perfection to those embracing Christianity or espousing the tenets of Neoplatonism. Medieval T-O maps and conceptual city maps already had appropriated this circular form. In the former, the world was depicted as a circle with a T within, the upper half representing Asia, the lower left Europe, the lower right Africa, with Jerusalem in the center. In the time of the crusaders, Jerusalem itself often appeared as a complete circle divided into quadrants.[28] Linked to ideals of the cosmos, the concept of Jerusalem as a sphere is related to that illumination of the earthly paradise in the *Très riches heures* of the duke of Berry, circa 1416. And it is this view of circularity that is transferred to images of Rome, whether in that same manuscript or in earlier fourteenth-century medallions, coins, and maps. Little evidence survives for classical antecedents, even in the maps of Rome, where rational proportional relationships appear to leap from the Several plan of A.D. 213 to Buffalini's map of 1551.[29]

Renaissance Transformations

In some ways, the more pictorial rendering of the eastern cities is explicable. City views are commonplace settings against which narratives are depicted from the mid-thirteenth century, whether in seals, manuscripts, or the trecento allegorical frescoes in Florence and Siena. Most notable of extant documents is the *Civitas Florentinae,* a detail of the fresco of the Madonna della Misericordia in the Sala del Consiglio of the Bigallo Oratorio. Dated 1352, it is the oldest known representation of Florence and characteristic of a

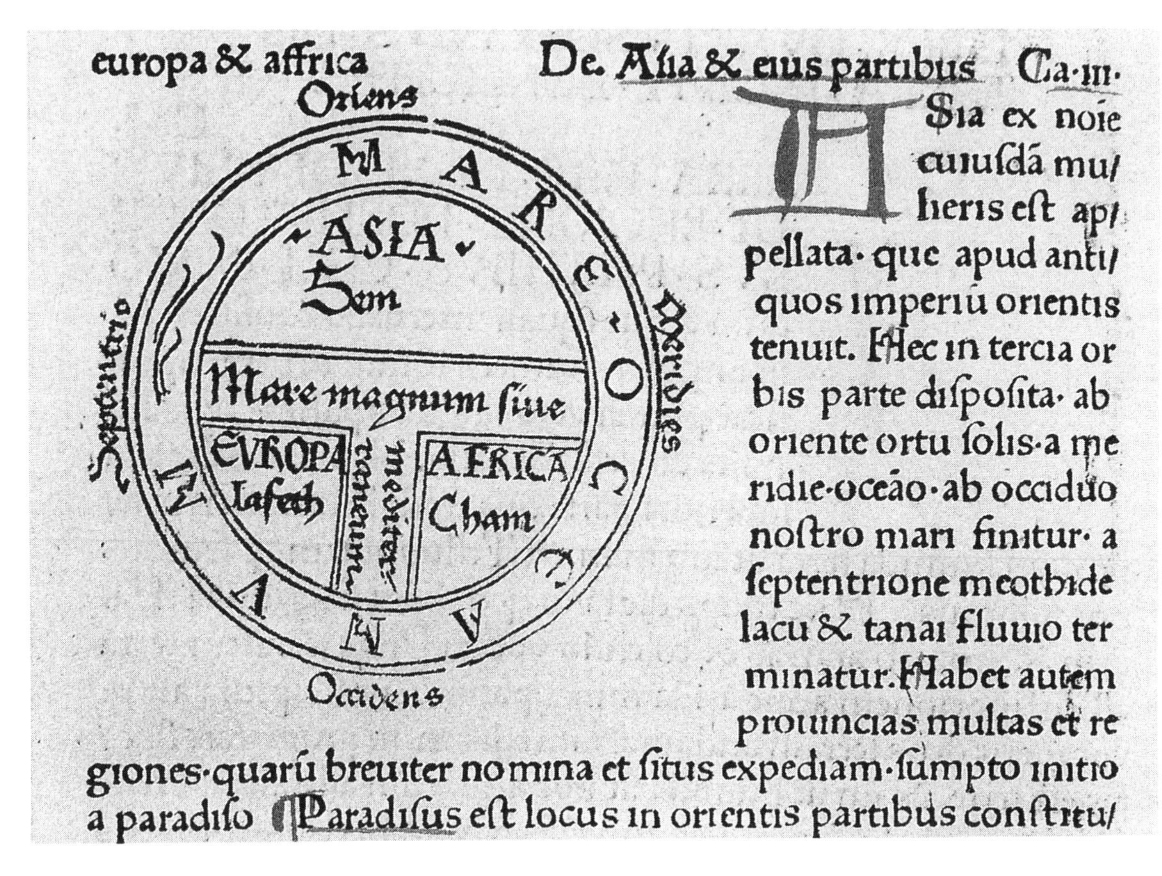

europa & affrica

De. Asia & eius partibus Ca·iii·

Asia ex noie cuiusdā mu/lieris est ap/pellata· que apud anti/quos imperiū orientis tenuit. Hec in tercia orbis parte disposita· ab oriente ortu solis·a meridie·oceāo·ab occiduo nostro mari finitur· a septentrione meotbide lacu & tanai fluuio terminatur. Habet autem prouincias multas et regiones·quarū breuiter nomina et situs expediam·sumpto initio a paradiso ¶Paradisus est locus in orientis partibus constitu/

FIGURE 2.8. T = O Map, the earliest printed map of Augsburg, 1472 (photo: The Newberry Library, Chicago)

genre in which the city appears in the palm of a saint or bishop, symbolic of its rule under divine guidance.[30] Here is the typical walled city with its cluster of towers and churches, densely packed within the urban fabric, as depicted in contemporary votive paintings.[31] In a similar genre is the quasi-topographical precocity in Giusto de' Menabuoi's above eye-level view "Padua in 1300" in the Luca Belludi Chapel of the Basilica del Santo. Painted in 1382, it depicts Saint Anthony assuming the protection of the city against a background of its monuments.[32] This penchant for an oblique setting cannot but recall trecento architectural compositions, such as Ambrogio Lorenzetti's "Allegory of Good Government" (1338–39), adorning the walls of the Palazzo Pubblico in Siena. Like the individual buildings in the Ptolemaic maps, the components of the fresco radiate outward from the center. John White notes that empirical perspective "is the pictorial counterpart of the spectator's situation, always looking outwards from himself, the center of his world. It is the reverse of the frozen stare of fifteenth-century perspective in which the composition is sucked in towards a single point by centering orthogonals."[33] Although both the city

FIGURE 2.9. Civitas Florentinae, detail of fresco *Madonna della Misericordia,* Loggia del Bigallo, Florence, 1352 (photo: New York, Scala/Art Resource)

FIGURE 2.10. Giusto de' Menabuoi, Padua in the year 1300, fresco, Chapel of Luca Belludi, Basilica del Santo, Padua (photo: Padua, Museo Civico)

maps and the painted city views are depicted without regard to mathematical principles of proportion, the latter do convey the impression of a unified entity, namely, a coherent panoramic townscape.

However, it is in the writings of Alberti that we find a new mode of cartography founded on scientific principles. Relying on a conception of the world based on a mathematical order, subject to certain measures and rules and to ratios in proportion, both the geography of Ptolemy and the art of the Renaissance succeeded in rendering forms in space with greater precision. Whereas Ptolemy had deemed it fit to compare aspects of geography with painting, it is Alberti who utilized the principles of Ptolemy's *Geography* for his own surveying and mapping. A similar "relational ordering of space" thereby unites the scientific and artistic revolutions heralding the modern worldview.[34] Like mapping, perspective focused on relative values regarding scale, size, shape, distance, direction, and, above all, on accessibility to the observer. Hence, Alberti's reference to a circumferator to plot locations by radial degrees and circular parallels as preparation for the first known scaled map of Rome in his *Descriptio Urbis Romae* (probably composed in the 1440s)[35] recalls the fixing of position using the latitude and longitude set by Ptolemy to measure the heavens in his text on astronomy.[36] For Alberti's perspective net as applied to painting is directly related to Ptolemy's system of parallels and meridians, based on a grid of polar coordinates that marked relative distances between specified

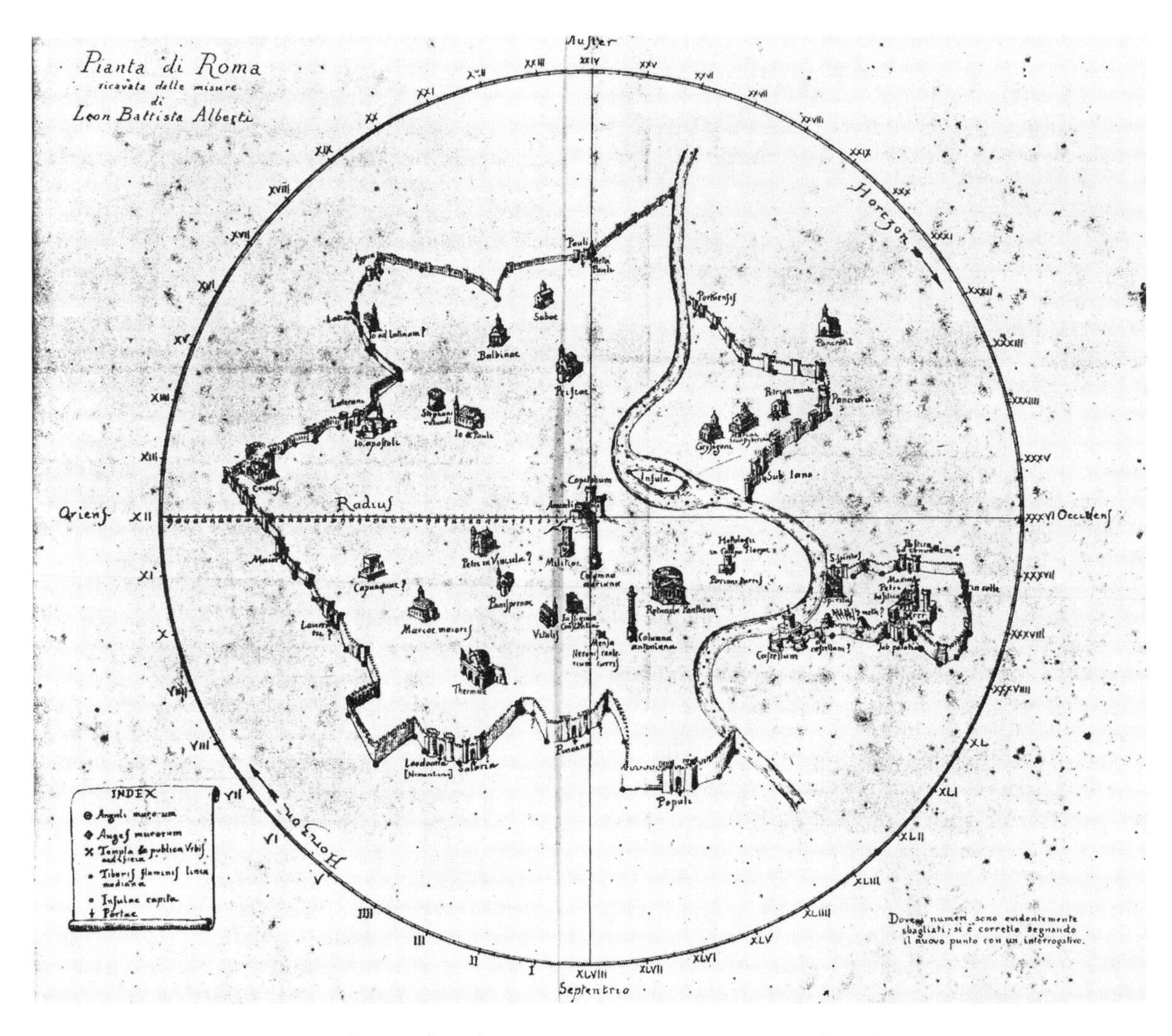

FIGURE 2.11. Leon Battista Alberti, Plan of Rome, ca. 1440s. Reconstruction by Alessandro Capannari, 1884 (photo: Rome, Istituto di Studi Romani)

points. Tied to the symbolic plans of Rome and Jerusalem in the Middle Ages, as also to the ideal Renaissance city, Alberti's use of the circular form enabled him to develop a coordinate system.

Alberti's map of Rome also harkened back to a system of ancient cartography, drawn from Greek principles of surveying, using instruments such as the astrolabe and quadrant and combining the coordinate tables of Ptolemy with the nautical models of mariners' charts.[37] His work was nurtured too by his contact with humanists, geographers, scientists, and artists who gathered in Florence and Rome in the 1430s and 1440s. Here was Flavio Biondo, Ciriaco d'Ancona, Poggio Bracciolini, Paolo Toscanelli, and Brunelleschi, expounding on the newly discovered classical texts. Here Ptolemy's *Geography* was discussed and eventually updated.

The *Descriptio,* however, may have been conceived during Alberti's first stay in Rome as papal legate from 1431–34, and formulated in the forties when he joined the learned circle that gathered around Pope Nicholas V. In the forties and fifties, Alberti's theoretical oeuvre was concerned with scientific exactitude and mathematical ratio. His discussion of the use of scaled relationships to reproduce objects in their just proportions had appeared in his first treatise, *De pictura* (1435). Further, his use of instruments in mapping the city reappears in *De statua* in his device, the *exempeda,* for measuring parts of the body: with it he calculated lengths, and with the movable squares, the diameters of the limbs.[38]

Massaio may have been aware of data used by Alberti, but only the maps of Florence and Rome appear to be distantly related to an Albertian modus operandi. Thus it appears that the Florentine mapmaker was not yet ready to apply the principles of scientific perspective to overhaul the standard representation of cities. Rather, he improved the rendering of buildings and clarified spatial relationships in accord with conventional bird's-eye views and oblique projection.

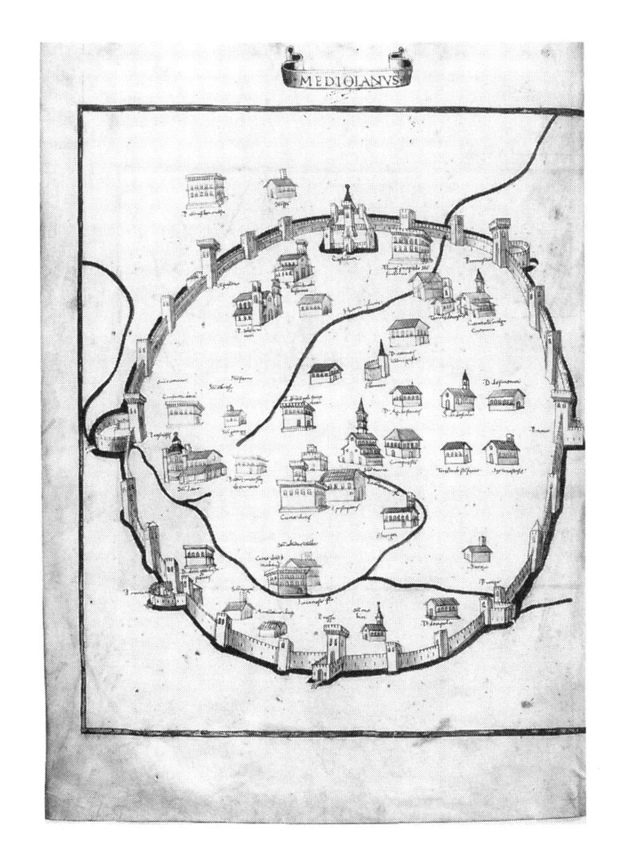

FIGURE 2.12. Milan, Plan view. From Ptolemy, *Geography,* Biblioteca Vaticana, urb. lat. 277, 1472, fol. 129v, Piero del Massaio (photo: Biblioteca Apostolica Vaticana)

City Maps: Description

TURNING TO THE INDIVIDUAL city maps, even a cursory examination provides a portrait of each city as seen through the eyes of the artist's contemporary Florentines. Details bring to the fore what is included or emphasized and what is excluded or minimized within the general parameters of city views cited above. Our study of comparative pictorial documentation and literary sources not only enables us to visualize the physical aspects of the city but also to enhance our understanding of its historical and cultural components. At the same time, we may achieve a deeper awareness of the imminent revolution in cartographic studies.

Milan

The form of the city is round in the manner of a circle, and such roundness is symbolic of its perfection.
—Bonvesin de la Ripa, *Grandezza di Milano,* 1288

Topographical features are most evident in this paradigmatic view of a circular walled city situated on a plain. Canals flow through the city to join the Ticino and the Po and extend beyond the walls. Sixteen symmetrically disposed quadrilateral towers and eight gates, including that of the Castello Sforzesco, form a radial plan, extant today. Northwest is at the summit, marked by the city's most important building complex, the Castello Sforzesco, begun by Galeazzo II Visconte in 1368–70. Within the walls, churches and palaces are casually disposed. Location is approximate but the search for a particular building is aided by sighting the nearest city gate. Witness the positions of Sant' Ambrogio near Porta Vigellina (Vergelina) and especially of San Lorenzo near Porta Temesis (Ticino); the latter's grand complex, with great blue dome and gilded cupola, colonnaded porch and campanile, is the dominant entity on the map. The Duomo, designated as Sancta Maria, is virtually at the center, its triangulation clear, its campanile alongside. Nearby, the palaces of the duke and that of the archbishop are prominent.

Even more marked are the buildings belonging to the patrician class, the quadrilateral two-story palaces, delineated by stringcourses; their upper stories with arched windows are crowned with heavy crenelated cornices and towers. As expected, the most conspicuous of these fortified Renaissance palaces is the Curia Ducis, just northeast of the Duomo. Two structures outside the walls at the top of the map summarize the dominant building types—the palace and the church, the Palazzo di Tommaso da Bologna and Santo Spirito.

Filarete's Ospedale Maggiore, begun in 1456, and a major building project at the time the maps were made, appears between the Porta Rosa and Porta Romana as Annunciata Hospitale.

Both the Urbino and Vatican maps were the subject of a detailed comparison by Achille Ratti (Pope Pius XI, 1922–39).[39] Like his late-thirteenth-century predecessor Bonvesin, Ratti begins his description quantitatively, citing eighteen churches in the Latin edition and seventeen in the Italian.

No discussion of Renaissance maps of Milan could omit mention of Filarete's ideal city Sforzinda with its circular walls and eight principal gates. Like quattrocento architects and Vitruvius and Ptolemy, Filarete based his plan and siting on the wind rose, a diagram comprised of four quadrants indicating directions of the winds.[40] Further, Filarete's attention in book 2 to building types and their location in the city, responding to hierarchical demands (church and government) and to exigency (the necessary markets and commercial structures), shares the spirit of contemporary mapmakers. His consideration of buildings in relation to each other and to the whole, both aesthetically and functionally, had a specific bearing on Renaissance ideas of town planning.[41]

Consider too the bird's-eye views of Milan by Leonardo da Vinci. Unlike his predecessors, who produced variations on the ideal plan focusing too on its symbolic connotations, Leonardo was more concerned with practical problems of water supply, maintenance, and circulation—that is, on those aspects constituting the infrastructure of the city.[42]

Venice

Venice is a very beautiful city with many inhabitants. It lies in the middle of the salt sea, without walls, and with many tidal canals flowing from the sea.

—Arnold von Harff, *Pilgrimage,* 1497

Like Venice itself, the plan is unique. This is, above all, a view of the lagoon, in which water is so pervasive that the separate districts of Venice appear but as larger, more compact islands and the city as a whole is but the largest in a sea of isles. However, if the artist shows appreciation for the queen of the Adriatic in its totality, he

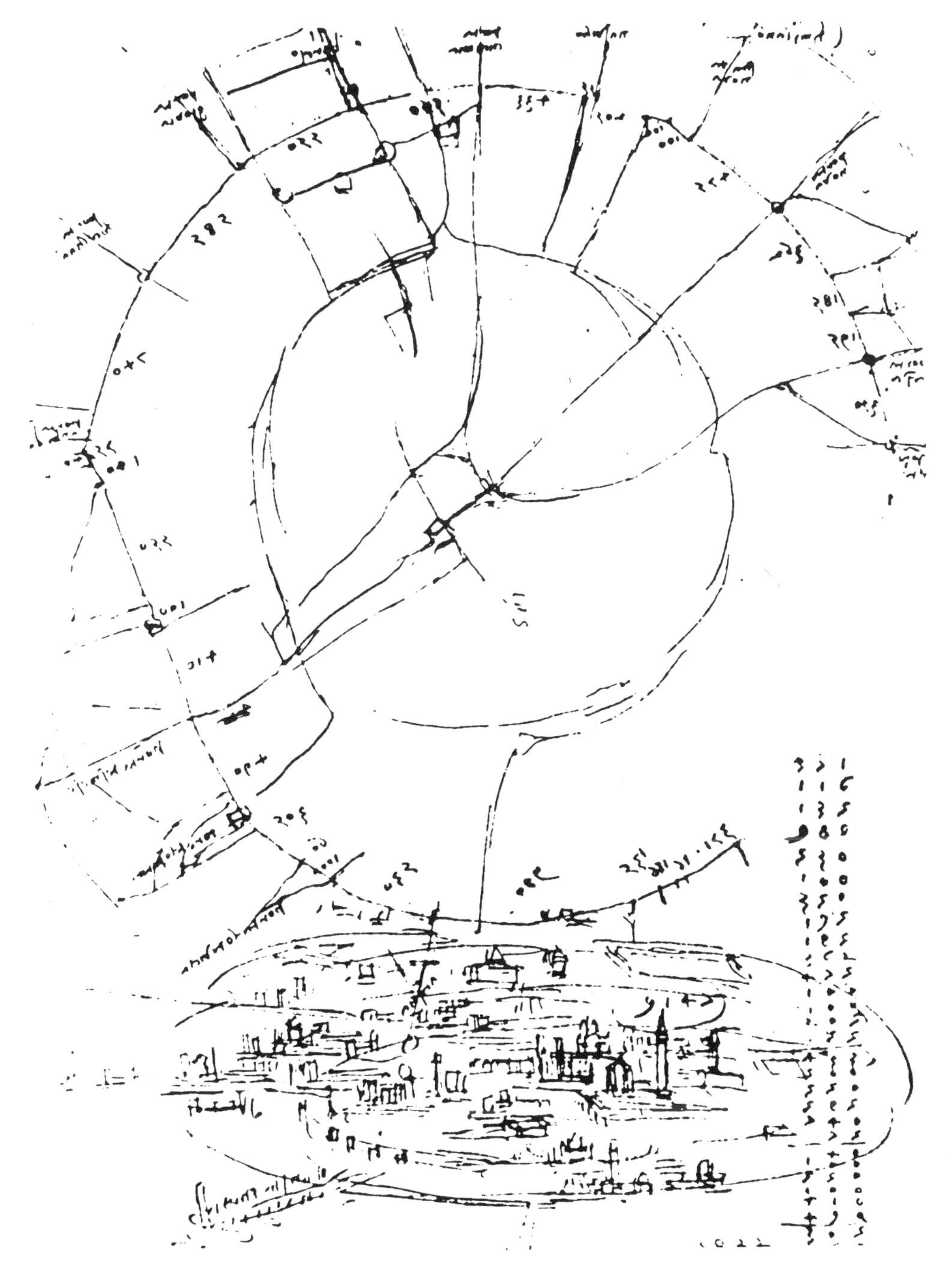

FIGURE 2.13. Leonardo da Vinci, Milan, hydraulic scheme, from Cod. Atlantico, fol. 199v, pen and wash (photo: Milan, Biblioteca Ambrosiana)

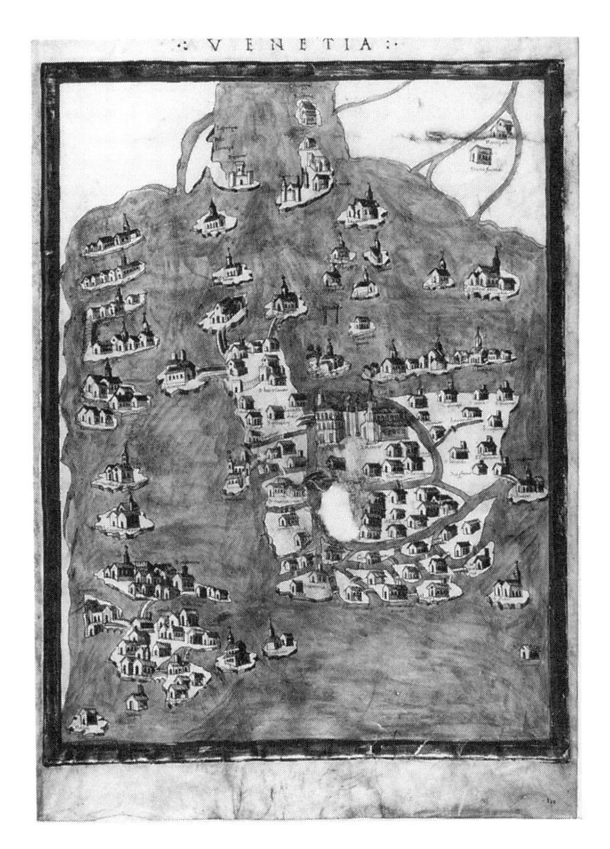

FIGURE 2.14. Venice, Plan view. From Ptolemy, *Geography,* Biblioteca Vaticana, urb. lat. 277, 1472, fol. 130r, Piero del Massaio (photo: Biblioteca Apostolica Vaticana)

presents little evidence for the late Byzantine and Gothic building fabric. Unlike the other Italian cities, emphasis on specific buildings is limited, whereas the overall geography is paramount. Each island is marked by a church or convent, though a few fortresses break this mode.

To my knowledge, the Ptolemaic maps of Venice are the only Venetian maps oriented with south at the top. Finding one's place within the city is relatively easy as the Piazza San Marco is clearly distinguished with San Marco, the Palazzo Ducale, and the two twelfth-century Veneto-Byzantine columns between, representing Saint Theodore and Saint Mark, the patron saints of Venice. The Bridge of Sighs is to the left of San Marco, and a two-storied arcade signifying the Palazzo Ducale reveals within an open court. Depicted here is a Venice crowned by cathedral, churches, and convents.

The group of islands constituting Murano to the north is readily identified, including its two bridges—one connecting to the south island where the Church of San Pietro Martire is visible. San Michele is correctly placed to the southwest, close to San Christoforo. Murano, San Michele, and San Christoforo are more or less accurately sited to the northwest, but the artist, perhaps due to a lack of preplanned space, relegated all the other northern islands to the west.

Near the Rialto is a building rivaling the Doges' palace in size and most crudely depicted. Is this building the Palazzo Camerlenghi, begun in 1488, a possibility seconded by the presence of the Naranzia lane alongside the facade? More probably, however, in this context, the large arcaded structure was the Fondaco dei Tedeschi, the location of the German merchants' warehouse from the early thirteenth century. For on this site near the Rialto there was a great confluence of traders and bankers from all Europe, with local Venetians catering to their demands. In 1473, the naval arsenal was greatly expanded with construction works in progress at the Darsena Nuova (Arsenale Novissimo).[43] Considering the rubric "arcana" on the Paris codex, is it this addition to the arsenal that was so roughly inserted in the Urbino manuscript?

Looking at the map of the city, we can now discern its division into *sestieri.* From left to right (or east to west) and top to bottom (or south to north) we may easily distinguish the following districts: Castello, Giudecca, San Marco, Dorsoduro, San Polo e di Santa Croce, and Cannaregio.

Despite its shortcomings and inaccuracies, and the arbitrary position of the islands and place names, the Ptolemaic maps are more than a compilation of monuments within the city. Practical considerations appear to be uppermost in the dominance of hydrographic elements and, perhaps, also in the map's adaptability to military purposes.[44]

In terms of the type of mapping it presents, the Ptolemaic bird's-eye view is closer to the earliest known surviving map of the city, circa 1330 (but available only in later copies), the parchment plan inserted in the manuscript recording the city's events compiled by Fra Paolino, the *Chronologia magna* of 1346.[45] Supposedly this map is based on a mid-twelfth-century prototype, drawn up by offices of the republic responsible for the maintenance, repair, and formation of the canals and islands.[46] Its most extraordinary feature is its consistent scale, raising the possibility of the utilization of measured surveys in the Middle Ages.[47] This tradition of official mapmaking is recorded too in 1460, when the Council of Ten decreed the drawing of local maps.[48]

Although separated by over a century, the Ptolemaic maps bear more than a superficial resemblance to the Paolino map, with its focus on navigation routes in the lagoon and the settled part of the mainland already in shape.

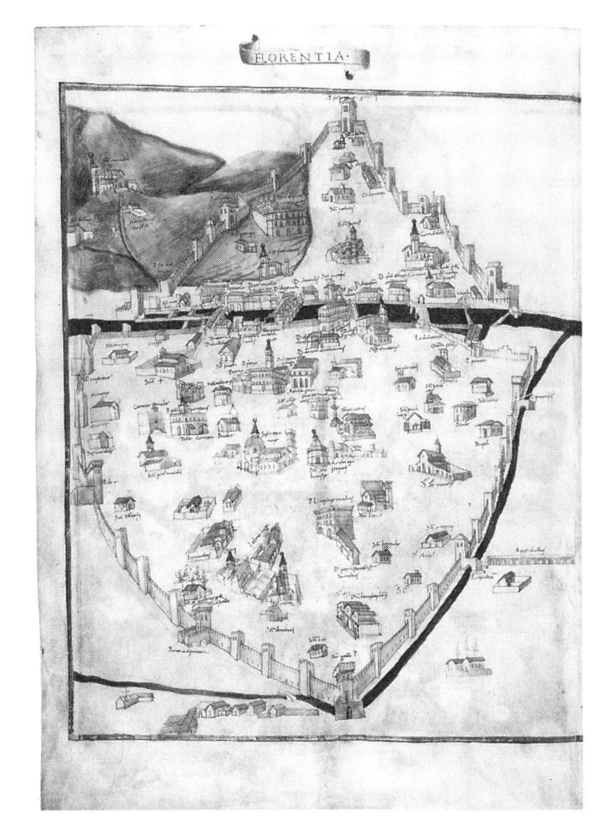

FIGURE 2.15. Florence, Plan view. From Ptolemy, *Geography*, Biblioteca Vaticana, urb. lat. 277, 1472, fol. 130v, Piero del Massaio (photo: Biblioteca Apostolica Vaticana)

Florence

But after I was brought back to this city of ours, adorned above all others . . . I realized that . . . there was talent for every noble thing not to be ranked below any who was ancient and famous in these arts.

—Leon Battista Alberti, *On Painting*, 1435

Not surprisingly, in terms of the provenance of the manuscript, the view of Florence is the most coherent and "correct" of the city maps. Dominated by churches and palaces, the city of the late Middle Ages is completely surrounded by walls, with its southern and northern parts bisected by the Arno River, here running a completely straight course. The Porta Gattolini (now Romana) to the south is at the apex. The Porta San Gallo, from which the Mungone River leaves the edge of the walls and flows through the landscape toward the west, is in the north at bottom. Most marked of topographical features are the hills on the southwest, dominated by the facade and campanile of San Miniato.

The overall shape of the map may be described as a pointed ellipse, with a strong east-west horizontal created by the Arno River and an almost visible south-north axis from the upper to the lower gates. Although streets are completely absent, they are implied by the location of the buildings.

Each building is seen as an isolated entity. These insulae are drawn at nearly the same scale—the Duomo (designated here as Sancta Reparata, the name of the earlier church on the site), a church among other churches, more prominent by virtue of its centrality than by its size. And, whereas a circulatory system of streets and roads is totally absent, buildings are placed on their approximate sites within the city in regard to the principal referent, the wall. Thus, a mere glance will reveal the Duomo, Baptistery, Palazzo Medici, Palazzo Vecchio, Loggia dei Lanzi, and Brunelleschi's recently built structures, the Ospedale degli Innocenti, Santo Spirito, and San Lorenzo.

Among the additions to the Paris map, none is more enhancing than the presence of gardens. Cypress trees rise from the cloisters of San Marco, Santo Spirito, and the Carmine, and gardens are visible adjacent to the grand palaces of the Pitti, the Medici, and in the Sapienza and the hospitals of the Innocenti and Bonifacio. Here there is a progression toward a townscape that points to its surrounding landscape.

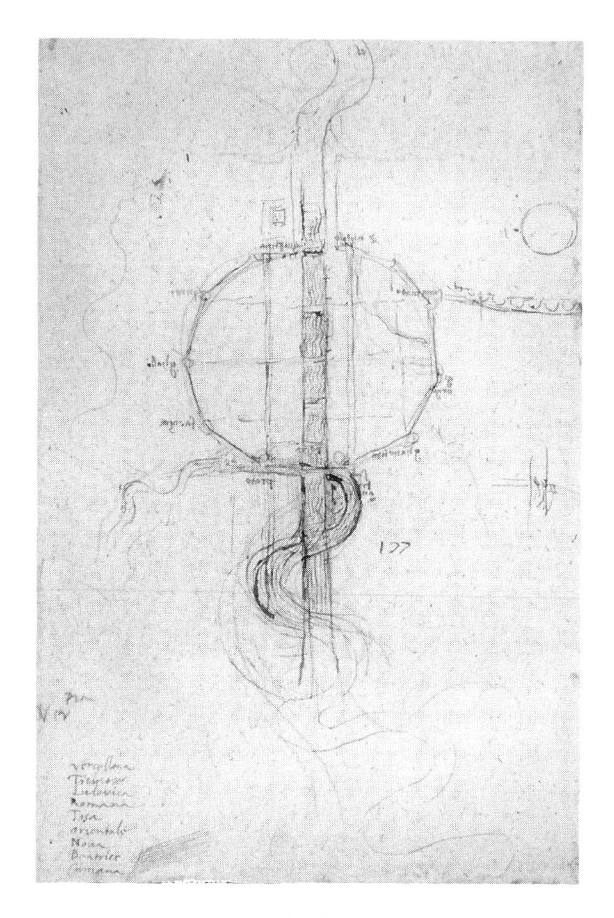

FIGURE 2.16. Leonardo da Vinci, Florence, Canalization scheme of the Arno. From Windsor, Cod. Atlantico, fol. 12681, pen over black chalk (photo: Reproduced by Gracious Permission of Her Majesty Queen Elizabeth II, Windsor Castle, Royal Library)

How may we situate the Ptolemaic maps within the corpus of Florentine city maps? Although remarkably little pictorial data exists, there is evidence for the existence of measured plans. A letter from Lapo da Castiglionchio, lawyer, records how Francesco da Barberino, judge, made a plan of Florence, now lost, dated circa 1325, which included "the entire city, all the walls and their dimensions, all the gates and their names, all the streets and piazzas and their names, all the houses that had gardens."[49] We are left, however, with a different type of documentation that provides some glimpse of the trecento city in the townscapes and architectural backgrounds of manuscript and decorative mural painting.

A fresco for the Guild of Judges and Notaries, dated 1366, is both a variant of a *mappamundi* and a reflection of Bruni's *Laudatio Florentinae Urbis*. Written in 1402, this encomium describes the city as the "geometric center of four concentric circles" and, within this nexus of power, the

Palazzo Vecchio is at the midpoint.[50] Although presenting a scheme for the hydraulic network of the Arno valley, Leonardo's polygonal map of Florence (Cod. Atl. 12681) evokes the circularity of Bruni's ideal city, nurtured by its Tuscan landscape.[51] Renaissance order has been imposed on the medieval city, while nature, represented by the Arno, flows freely beyond the city walls.

Rome

Together [*Brunelleschi and Donatello*] *made drawings of all the buildings in Rome . . . with the measurements of the width, length and height, so far as they were able to ascertain them by judgment.*

—Antonio Manetti, *Life of Filipo di Ser Brunelleschо,* circa 1480

Oriented with south at the top, the map follows the course of the Tiber with its island further establishing our bearings. On both sides of the river, two elevated masses dominate Rome's famous hills. Whereas classical and archaeological monuments are dominant, religious edifices are hardly lacking, and a few sites even record the Christian events with which they were associated. But not a single private palace is visible. Although aqueducts are clearly discernible, streets are completely absent, except for the Via Triumphalis leading from the Castel San Angelo to the Capitoline.

The Urbino map, the most detailed of the Ptolemaic views, presents the greatest concentration of buildings in the Vatican area and around the adjoining Gianicolo and Aurelian walls. Renowned monuments, such as the Colosseum, Pantheon, and Thermae, are portrayed with remarkable lucidity. Almost at the very center is the Capitoline Hill, its steps leading to Santa Maria in Aracoeli and the adjacent Tabularium. Drawings of the Dioscuri and equestrian Marcus Aurelius show these statues in their old locations, the former on the Quirinale and the latter

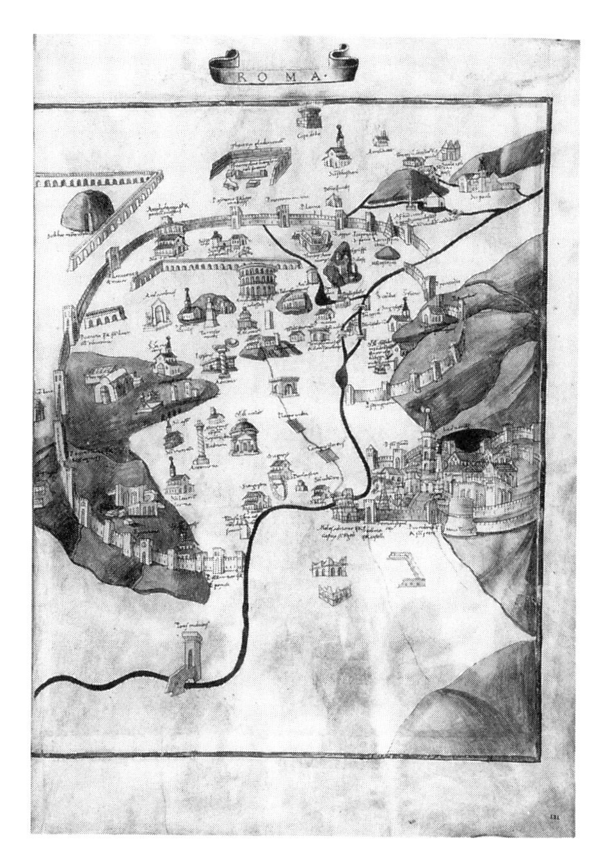

FIGURE 2.17. Rome, Plan view. From Ptolemy, *Geography,* Biblioteca Vaticana, urb. lat. 277, 1472, fol. 131r, Piero del Massaio (photo: Biblioteca Apostolica Vaticana)

before the Lateran (from whence it was transferred to the Capitoline in 1538). Ruins of ancient walls are clearly visible outside the boundaries of the city, as is the Ponte Milvio standing isolated at the north end. In front of Santa Maria in Trastevere, the old *rione* fountain bears the legend "unde oleum fluxit usque Tiberim nocte Nativitatis Domini" (where oil flowed to the Tiber announcing the birth of Christ). This rubric provides a strong indication that the origin of the Ptolemaic maps lies in the *mirabilia* type of pilgrimage guide. However, rather than suggesting a special emphasis on the Christian aspect of the city within a larger archaeological and antiquarian context, the density around the Vatican is symptomatic of the stepped-up building activity then in progress.

To some extent, Massaio's maps must reflect the vast construction program undertaken by Nicholas V (1447–55), the restoration of the walls, and, above all, the transformation of the Borgo and the Vatican Palace and the rebuilding of Saint Peter's. In the rendering of the Vatican area, it is apparent that Massaio has brought his "unknown" prototype (probably circa 1404–20) up-to-date by including the tower of Nicholas V, completed in 1454, inscribed "Nova turris."[52] Further, the double circle of walls that rings the Vatican complex, no longer extant in 1472, appears to be an addition of the artist (or a direct copy from an earlier model), as it is absent on earlier maps.

Among all the maps related to those of the Ptolemy manuscripts, none is more significant than the so-called Strozzi plan of Rome, dated 1474. The possibility that the Ptolemaic maps are reduced variations of this map is due to common topographical features, similar renderings of principal monuments, and especially certain shared idiosyncrasies. While conceding the use of *Mirabilia* literature as one source for the sites of legends and relics depicted and the histories of known places, the Strozzi map is probably tied to Flavius Biondo's *Roma instaurata,* composed in 1444–46.[53] Analogies appear in the panoramic view, the disposition of walls, the use of both ancient and medieval sources, and the siting of landmarks. Moreover, Biondo's work evolved from archaeological studies by humanists and their foundations in classical texts; its topographical classification of buildings according to types was constantly updated by the additions of new construction, such as that initiated by Nicholas V in 1451–55.[54]

Yet, no source appears to rival that of Alberti's projected map of Rome as the principal model for both the Strozzi and Ptolemaic views. The key to this plan is in the *Ludi Matematici,* written circa 1450, wherein he discusses the use of polar coordinates to measure the circumference of a territory, that is, the construction of a city plan on a mathematical basis.[55] Even though Alberti's goals were not completely realized—for his plan lacks streets, height limits, and land areas—his method of plotting distances in a scientific manner represents a watershed in mapping.

Constantinople

A city which was so flourishing, with such a great empire, so many illustrious men . . . so prosperous . . . she who was the head of all Greece, the splendor and glory of the East . . . has been captured, despoiled, ravaged, and completely sacked by the most inhuman barbarians and the most savage enemies of the Christian faith, by the fiercest of wild beasts!

—Cardinal Bessarion, 1453[56]

Familiar in the annals of cartography, the Ptolemaic views of Constantinople are based on a well-known, often reproduced early fifteenth-century map by the Florentine priest Christoforo Buondelmonte, which appeared in his *Liber insularum Archipelagi* (1422). Here is a map of the great Eastern capital until its conquest by the Turks.

Oriented with Pera (also known as Galata) at the top, the peninsula is surrounded by the Golden Horn on the north and west, the Sea of Marmora on the south, and the Bosphorus on the east. The deep harbor and multiturreted, strongly fortified walls enclose a city that still bespeaks its Byzantine splendor in the years preceding its rebuilding under the Ottomans.[57] To-

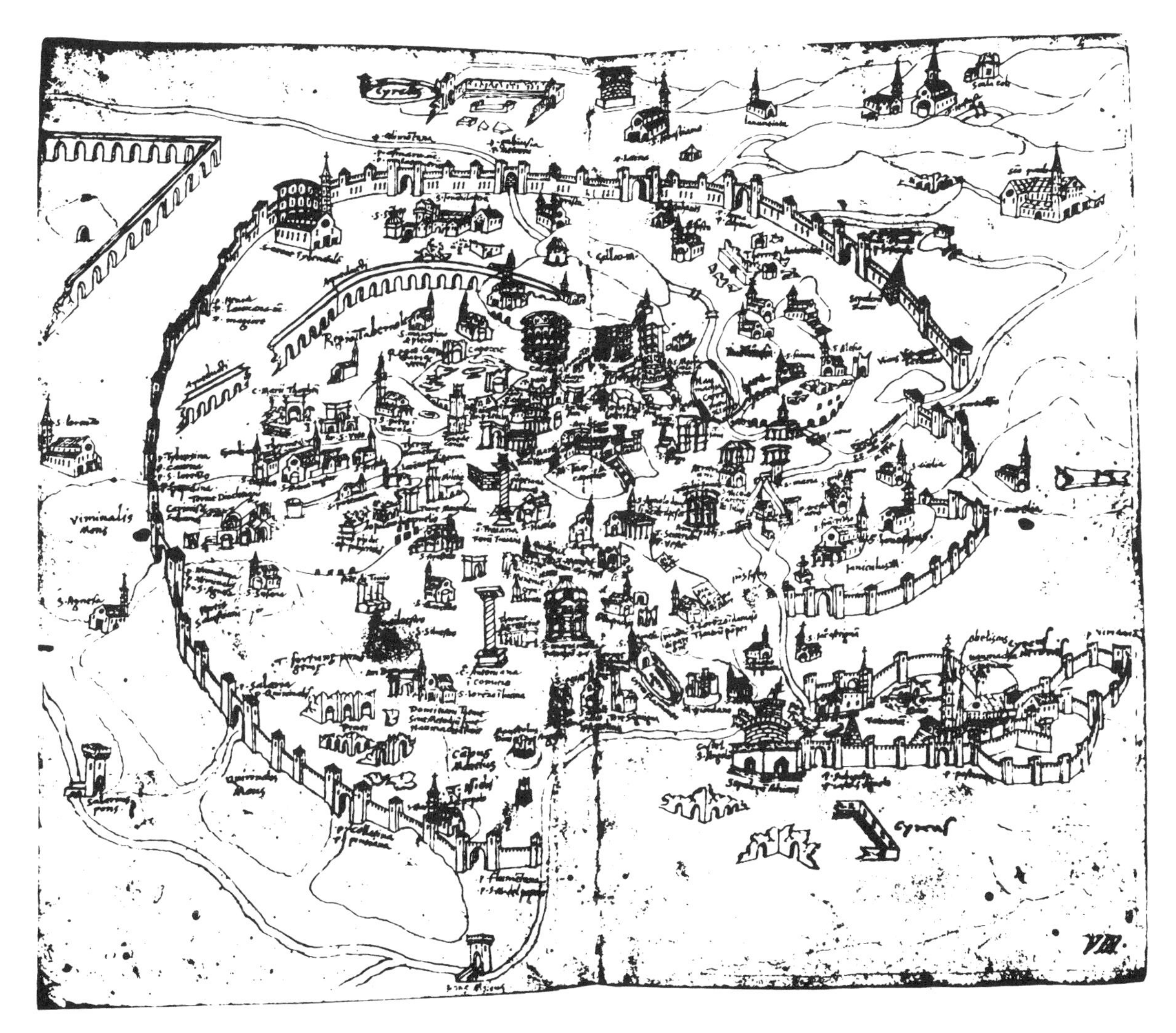

FIGURE 2.18. Alessandro Strozzi, Rome, Map, pen drawing, 1474. From Biblioteca Medicea Laurenziana, Laurenziana Redi 77, fols. viiv–viiir (photo: Florence, Biblioteca Laurenziana)

ward the north, the walls and towers of Pera are visible, along with a single windmill. If the humanist scholars looked to Constantinople as a font of ancient wisdom and embalmed classicism with its storehouse of Greek manuscripts, Florentine merchants lodged in Pera's thriving Genoese colony undoubtedly knew it as the site of one of the richest alum mines in Asia Minor.[58]

Within the walls, the massive three-tiered circular form crowning the eastward hill of the city is Hagia Sophia. Its arched openings, blue dome, and gilded lantern are engulfed within the walls

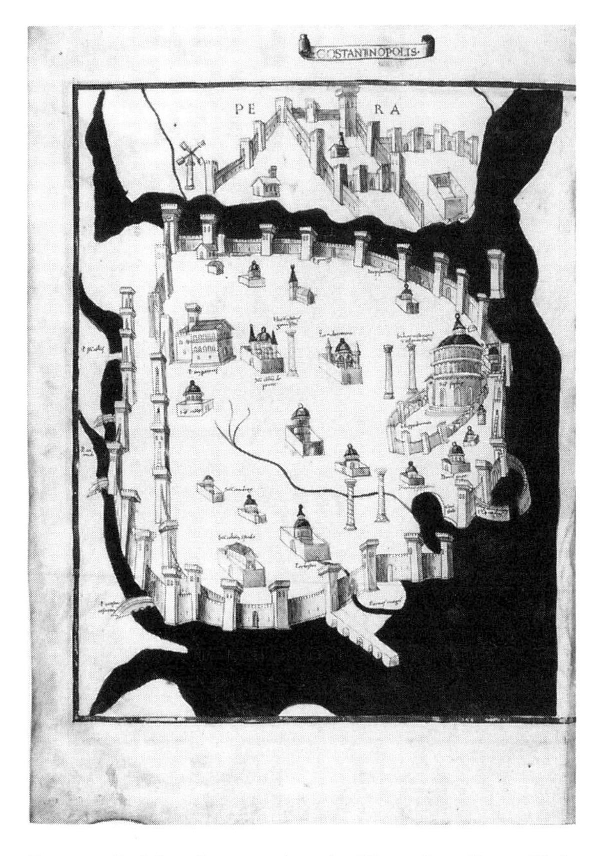

FIGURE 2.19. Constantinople, Plan view. From Ptolemy, *Geography,* Biblioteca Vaticana, urb. lat. 277, 1472, fol. 131v, Piero del Massaio (photo: Biblioteca Apostolica Vaticana)

of the great Hippodrome, the site of the emperor's epiphany, which, together with the city walls, received top priority in Constantine's building program.[59] The churches of Saint John in Studium and Saint John in Petra are depicted as circular edifices with arched openings and blue domes; crowned by gilded spheres and accompanied by minarets, they are set within quadratic compounds. Some important monuments are noticeably absent, namely, the Church of the Apostles (destroyed in 1463), Saint Irene, the gate of the Jews, and the imperial palace. Only

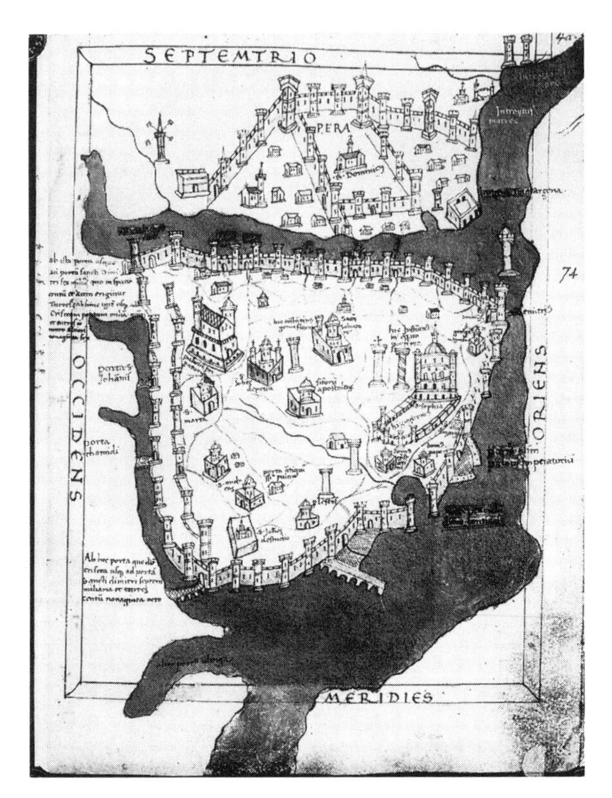

FIGURE 2.20. Buondelmonti, Constantinople, Map. From Biblioteca Medicea Laurenziana, Laurenziana XXIX.25 (photo: Florence, Biblioteca Laurenziana)

two of the five famous columns on the map are identified, namely, the column of the cross where Constantine kneeled, and that of Justinian, once adorned with a bronze equestrian statue of the emperor. These monuments are described in Buondelmonti's widely disseminated "island" book, which provides a remarkable document expressing the scholarly fervor for geographical, archaeological, and antiquarian data in the early fifteenth century.[60] Considering the gatherings of humanists, mathematicians, and explorers who exchanged geographic information in Florence at this time, it is hardly surprising that the many copies of Buondelmonti's manuscript and views of Constantinople provide us with vivid portraits of the city on the Golden Horn. They not only reveal the city's topography, its walls, and its notable structures, but they also perpetuate the myth of Byzantine dominance. For, conforming to its prototype, Massaio's maps depict the Christian city with its churches and campaniles rather than a landscape in a state of Islamic transformation.

Jerusalem

> *. . . and* [*the angel*] *showed me the holy city of Jerusalem coming down out of heaven from God. . . . The city lies foursquare, its length the same as its breadth.*[61]
>
> —*Revelation* 21:10, 17

Jerusalem, the ultimate pilgrimage city, is here depicted on a rocky ridge between the Kidron and Hinnon valleys, its legendary fortified walls in a ruinous state, true to their condition in the fifteenth century. Oriented toward the west, this map shows the Christian buildings of the walled city from the Mount of Olives. The group dominated by the Church of the Holy Sepulchre is immediately visible as the true sign of the city.

For all its crudeness, the Ptolemaic map provides a coherent view of the city as a whole and serves as a practical guide to the holy places. Even simplified, this map conveys some of the compactness and range of the structures and sites of important biblical events, from streams and fountains to tombs and altars in a city where every stone and field were associated with a passage from the Scriptures or a legendary act.[62] Eyewitness accounts thus fused with biblical sources and sacred texts as well as the writings of the church fathers.[63] Still, it is amazing that the roughly quadrangular shape of the rock-hewn mount of the Holy City and the general placement of the principal buildings within would be recognizable to a modern tourist. On

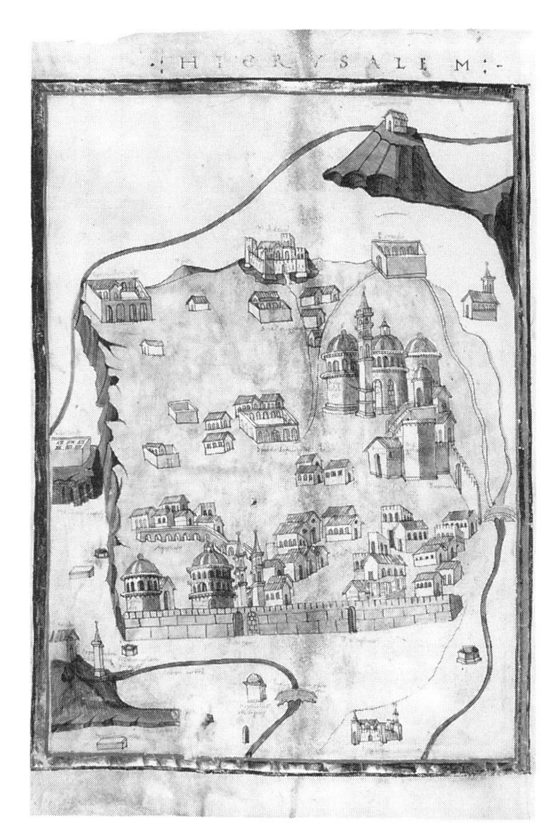

FIGURE 2.21. Jerusalem, Plan view. From Ptolemy, *Geography*, Biblioteca Vaticana, urb. lat. 277, 1472, fol. 132r, Piero del Massaio (photo: Biblioteca Apostolica Vaticana)

the east, the valley of Jehosophat, the Kidron stream, and the places cited at the foot of the Mount of Olives are compatible with the actual sites.

Emphasis remains on those events connected with the life of Christ and the Virgin, as well as on the actual shrines and goals of Christian pilgrims. Some of the monuments are described by the events that they harbored, such as Christ's Last Supper. Noteworthy, too, outside the walls toward the Mount of Olives is the site of the Garden of Gethsemane and the Virgin's tomb. Crossing the Kidron stream is a bridge, whose inscription states that the crucifix was made from its wood.

Just outside the southern wall toward the Valley of Hinnon, a type of rock-cut tomb, a horizontal catacomb-like mass, marks the pilgrims' cemetery built by the knights of Saint John in the mid-twelfth century.[64] Known as the Hakeldama or Field of Blood, this was in ancient days the site of the potters' field that served as a burial ground for foreigners.[65]

Massaio's map with the walls in a state of disrepair is a record of the city prior to their restoration from 1537 to 1549, and to the opening of the new gates by the Ottoman Sultan Suleiman.[66] Here, notable on the eastern wall is the presence of the gate now called Saint Stephen and, further, the Golden Gate, with one of its two doors open. Within the walls, on the earth-brown ramparts of the fortified city, the Holy Sepulchre and its tall quadrilateral campanile seem immense.[67] In the triad of centrally planned blue-domed buildings, the middle one is distinguished by its slightly greater breadth, arched openings, and lantern. Left to right, these structures are designated as "holy Sepulchre" (the Anastasis rotunda), "Mundi medium" (middle of the world), and Monte Calvario (the Mount of Calvary).

At the western wall stands the citadel of David, a mighty fortress with its prominent tower, the whole surrounded by a moat. Just below is the Church of Saint James at the site of his martyrdom. Conspicuously situated like an island at the center of the map is the hostel for pilgrims.[68]

Damascus

Damascus is a large royal city. . . . It stands in a broad plain, and is surrounded by a wide circuit of wall, which is fortified with closely shaped towers. Around the outside of the walls it has many olive-groves and the four large pleasant rivers which run by make it wonderfully fertile.

—Arculf, "Pilgrimage," ca. 679–88[69]

The view of Damascus may be more readily classified as a townscape than a map. In comparison with the other maps, its most extraordinary feature is the difference in the scale of the buildings represented. We see the old walled city, across whose northern boundary runs the Barada River and beneath the hills to the north the A'waj (Pharphar) River and their tributaries. Surveying the vista at the top of the map, oriented toward the south, is the Tower of Lebanon, from whose height one could discern the range of snow-covered peaks of Mount Hermon. Although a street pattern is absent, the Derb el-Mustakim or the Street called Straight running from the eastern to the western gates may be discerned.

Examining the spatial relationships between the buildings and their environment, we detect a certain coherence of another order. The city is no longer an aggregate of monuments but rather is viewed as a policentric entity. One may readily distinguish different architectural features—the tiles of the centrally planned domed edifices, the open loggias of the madrashes, the red topped roofs of the more commonplace vernacular.[70]

Despite the sophistication of this city map, it seems to be rooted in the literature of pilgrims, crusaders, and merchants, for two noteworthy events connected with the city are commemorated outside the walls. Paul's miraculous vision en route to Damascus is represented by the house from which he escaped, visible to the southwest, beneath the Tower of Lebanon. The other significant monument is the mountain to the north of the city, with its shrine marking the alleged spot where Cain killed his brother Abel.

FIGURE 2.22. Damascus, Plan view. From Ptolemy, *Geography*, Biblioteca Vaticana, urb. lat. 277, 1472, fol. 132v, Piero del Massaio (photo: Biblioteca Apostolica Vaticana)

Within the walls, the view is dominated by two buildings: one, a centrally planned structure with a cupola-crowned dome on a high drum, and the other, a massive basilica-like edifice with three tall quadratic towers, surmounted by graceful, orb-topped pyramids. The latter, extending from the northern wall, is the Great Mosque built in A.D. 705 on the site of the Christian basilica dedicated to Saint John the Baptist and supposedly harboring his head as a sacred relic.[71]

Yet this view hardly constitutes a map at all. Aside from the topographical features, little of ancient, medieval, or modern Damascus is represented. Indeed, in the midst of an exotic landscape, we see an oriental and insular walled city, but upon closer inspection, buildings of Roman antiquity coexist with those of Renaissance Florence. Above all, we are struck by the predominance of domed centrally planned buildings, a type of Templum Domini, representing the Dome of the Rock in the Holy City.

The city in the Ptolemaic map is girded by its fortification wall rebuilt in the twelfth century and betrays no inkling of the vast devastations heaped upon it by the Tartars in 1300 and by Tamerlane a century later. By the mid-fifteenth century, in the wake of four hundred years of invasions, the great caravan trade of Damascus had precipitously declined.

Alexandria

Alexandria the metropolis of all Egypt.

—Ptolemy, *Geography*

Oriented with south on top, the richly colored view depicts a walled city by the sea, reminiscent of some mid-trecento Sienese paintings. Hence, the site of the lighthouse, one of the seven wonders of the ancient world, appears below the entry at the eastern or Great Harbor. In the ancient Eunostos, or western harbor, stands another lighthouse. An unidentified building, visible a short distance from the site of the *pharos*, represents the Temple of Isis on the isle of Pharos.[72] Guarding the entry to the Great Harbor, the cotton warehouses, customs, and arsenal enclose a large pool, perhaps the location of the old royal port. Toward the north harbor, where once stood the greatest library of antiquity and the famous museum, there is a relatively dense aggregate of churches, mosques, and domes. Within the

FIGURE 2.23. Alexandria, Plan view. From Ptolemy, *Geography,* Biblioteca Vaticana, urb. lat. 277, 1472, fol. 133r, Piero del Massaio (photo: Biblioteca Apostolica Vaticana)

walls, the old Canopic Street is almost distinguishable with its tombs and mosques leading to the rocky outcropping at the eastern wall. This may represent the ancient Paneum (the sanctuary of Pan), the artificial hill just south of the tomb of Alexander, near the center of the city.[73]

Buildings depicted in the southeastern quarter appear to be part of the Arab city that rose on ancient ruins, wherein one finds the mosque of the Prophet Daniel. Two Byzantine structures are particularly intriguing, for their tripartite convex roofs seem to be vaulted with wooden ribs. Could they possibly depict the wondrous temple dedicated to Arsinoe, the Cyprian Aphrodite and maritime goddess, built by the admiral Callicrates and allegedly vaulted with elliptical arches?[74]

Rising above the city in the southwest is the great natural outcropping of rock (later much enlarged), the pre-Alexandrian ancient citadel of Rhacotis, where Osiris was worshiped before being replaced by the Greeks with Serapis.[75] Clearly marked on the map north and west of the shrine is the so-called Pillar of Pompey, as dubbed by the crusaders, but actually a column dedicated to Diocletian.

In the southeast part of the map, there is a mosque whose Byzantine architecture is quite distinct. Close by are cubic habitats with open upper loggias beneath flat roofs. This type, prominent in the map of Cairo as well, was undoubtedly suitable to the extremes of local climate. Outside the walled city, legend seems to predominate in the chapels or tombs of Saint Mark (who is reputed to have brought Christianity to Alexandria and from whence his body was transported to Venice in A.D. 828) and Saint Catherine.[76]

Most provocative of all, considering the date of the Urbino codex, is the depiction of the *pharos* on the eastern end of the island. In place of the fifteenth-century marble ruins of the lighthouse, exacerbated by earthquakes in the years A.D. 956, 1303, and 1323, we see the Fort of the Mameluke Sultan Qait Bey, built as a defense against the Turks in 1480.[77]

On the Ptolemaic map we do not see the city of the Ptolemies laid out by Dinocrates under Alexander the Great (332–331 B.C.) on a Hippodamian grid, its garden suburbs to the east, its hugh necropolis to the west. Not even the two most prominent landmarks, the obelisks from

Heliopolis known as Cleopatra's needles, located in the Caesareum, are visible. Nothing of the orthogonal street plan is suggested, except for the rectilinear outline of the walls, reinforced by regularly spaced towers.[78] Not unexpectedly, maritime features are pronounced, as is the ancient division into the two great eastern and western harbors.

Appropriately, the map may be viewed too as bestowing homage on Ptolemy's birthplace, and as a memento of its famed institutions of learning.

Cairo

> *Cairo of Babylon is a big city, and is all crowded with buildings: . . . [it] is so great that a courier cannot round this city in two days.*
>
> —Fra Niccolo di Poggibonsi, *A Voyage beyond the Seas,* 1346–50

At first glance, the lack of walls completely enclosing the city, together with the excessive whiteness of the folio, distinguish Cairo from the previous city views. A relatively short walled segment with four turrets and an arched entry—the remains of the eleventh-century Fatimid walls toward the northeast—appears between two of the tributaries of the Nile that flow through the city. Only two sites are identified on the Urbino map, which is oriented with south on top—the Castello del Soldano and the Umea balsami.

Perched on a rocky promontory, the Maqattam hills, at the southeastern part of the old city, the Castello del Soldano represents the great citadel, the sultan's palace, begun in 1176 by Saladin. It is seen here as a quadrangular fort from whose ramparts project a domed tower and whose sides are aligned with minaret-like turrets.[79] Part of the defensive bulwark, the citadel's summit commanded a sweeping panorama of the entire city.[80] Five mosques are within the cit-

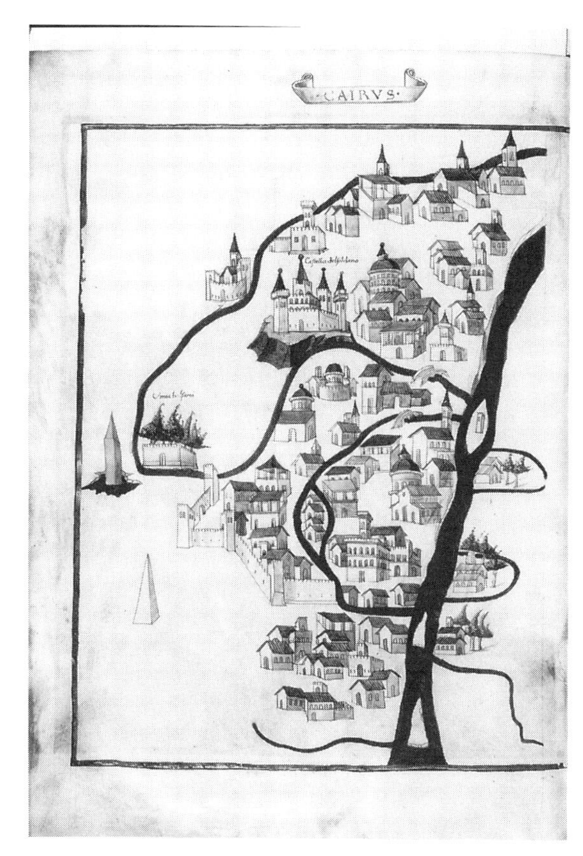

FIGURE 2.24. Cairo, Plan view. From Ptolemy, *Geography,* Biblioteca Vaticana, urb. lat. 277, 1472, fol. 133v, Piero del Massaio (photo: Biblioteca Apostolica Vaticana)

adel's walls, and to the west an aggregate of structures includes such notable Islamic landmarks as Ibn Tulun, the oldest mosque in the city, dating from the ninth century. Considering the vivid, extant fifteenth-century descriptions of an exotic, oriental city, it is obvious that the indigenous architecture has been westernized.[81]

The "Umea balsami" inscription appears above a square walled enclosure with ample balsam trees protruding. Here in the Balsam Garden the Virgin supposedly rested on her flight from

Herod's persecution, and here her famed tree flourished, allegedly on the same site where a tree dedicated to the goddess Isis once grew.[82]

Arbors appear too within the tributaries of the Nile and mark the suburban quarters. Within the city there are eminent palaces, such as the one noticeably Florentine edifice, a variation of the Palazzo Vecchio. It stands alongside domed religious buildings with lanterns topped by orbs, and sacred structures marked by pyramidal pointed towers crowned with spheres. A few houses depart from the Renaissance norm and show upper loggias with flat roofs, their interiors open to prevailing winds. Hence we see some attention given to the local vernacular, habitations comparable to the madrasah (the Islamic religious school), or inner courtyard, and compatible with climatic conditions.[83]

Beyond one of the Nile's tributaries at the far east of the map is an obelisk-like structure rising from a body of water. Despite its location here, it conforms to the description of the nilometer, the octagonal pillar erected above a square well to measure the annual inundation of the river. The oldest one, built in A.D. 716 on the southern end of the island of Roda, may here be represented by the quadrilateral turret on the island in the midst of the Nile, situated where Geziret Bulak is now.[84] A lone pyramid at the northeast of the map is the most obvious reminder to the modern viewer of ancient Egypt.[85]

Volterra

Volterra—up on a great mountain is strong and ancient—as is no other place in Tuscany.

—Fazio degli Uberti, thirteenth century[86]

Few perusers of this manuscript of Ptolemy's *Geography* would be prepared for its final folios. Another cartographic mode appears as we conclude our journey through this volume, the bifolium of Volterra rendered in a manner entirely different from that of the preceding maps. Inserted in the Urbino codex at a later date, the map must commemorate the capture of the city by the Florentines in 1472, under an army led by the duke of Urbino. In terms of contemporary events, evidence such as the position of cannons and the prominence of fountains and water sources points to the use of this map's model for strategic purposes. This is not a city map nor even a city panorama, but instead is more akin to a Renaissance painting of a town and its territory. At the same time, it is a genuine portrait of the city and its surroundings, documenting its contemporary appearance. Rather than being based on a copy of an older prototype, this view of Volterra, the ancient Etruscan city southwest of Florence, responds to the landscape's actual situation and the town as portrayed in a late quattrocento manner, employing aerial as well as parallel perspective and fully cognizant of a ground plan. Even the place names are popularized, written in the vernacular rather than in Latin as on the other maps.

Instead of a series of isolated monuments which define most of the city views, here the principal buildings and the ordinary domestic architecture are placed in relation to each other and, what is more, to the broad hilly landscape outside the walls. Topographical features are uppermost and, to a great extent, determine the location of buildings as well as fortifications. Cliff erosion affects the site of the Badia; the mills are of course along the streams; the fountains throughout also serve as cisterns. All the details requisite for a military campaign are here: witness the cannon on the natural fortress composed of rocky mounds in the upper left, and their inscription "flight from S. Andrea."[87] And, indeed, in the account of Baldi, we note that San Andrea occupied a strategic position during the

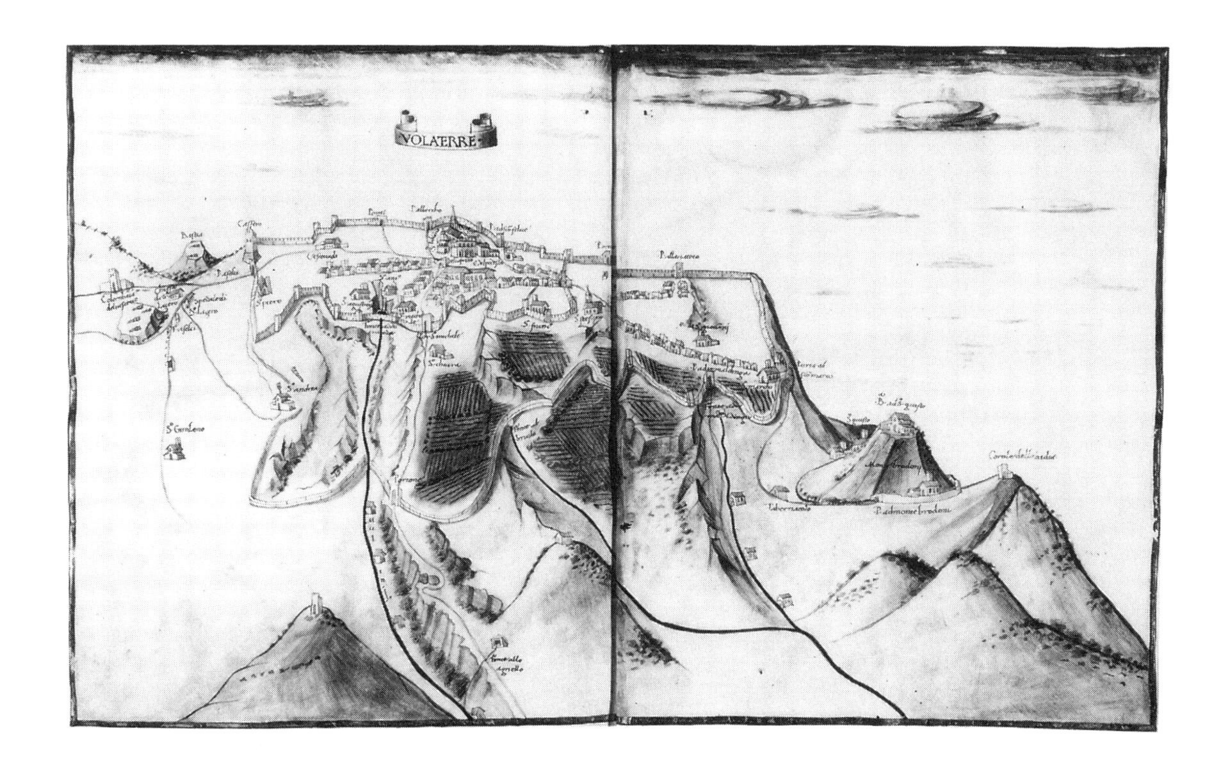

FIGURE 2.25. Volterra, Plan view. From Ptolemy, *Geography,* Biblioteca Vaticana, urb. lat. 277, 1472, fols. 134v–135r, Piero del Massaio (photo: Biblioteca Apostolica Vaticana)

siege, for being on the very outskirts of town, near an opening in the walls, it provided an entry point for the soldiers into the city.[88] Benedetto Dei in his chronicle for the year 1472 cites the bombs and ammunition placed where "the most worthy captain had ordered."[89]

Federigo's return to Florence was triumphal, and to him Jacopo Bracciolini dedicated the manuscript of his father's *Storia di Firenze,* 1472 (*Vat. Urb. lat. 491*), originally written in 1455. Appended to the manuscript, the equestrian portrait of Federigo bedecked in laurel appears before a panorama of the besieged city, its walls and buildings easily discernible from the Urbino map.[90]

In one folio, we have moved from the medieval map as a collection of isolated phenomena—admittedly, often in relation to each other—to a unified view of the city characteristic of Renaissance art.

CONCLUSION

IMAGES OF CITIES are prominent among works of art in the Renaissance, maps being part of the same phenomenon of artistic creation. Maps reflect too the growth of the city-states in many facets and, not least, the establishment of new civic rights and responsibilities. Generated by the rise of the communes, the town views parallel literary texts in praise of the city. Exemplified in Leonardo Bruni's *Laudatio urbis Florentinae,* this rhetorical trope is part of the *laudes civitatum* hon-

FIGURE 2.26. Duke Federigo da Montefeltro, equestrian portrait with Volterra in the distance. From Poggio Bracciolini, *Storia di Firenze,* 1472, urb. lat. 491, fol. IIv (photo: Biblioteca Apostolica Vaticana)

oring "the dignity of the founder . . . the aspect of the walls, the region, and the site."[91] Like the texts praising moral and political superiority and good government, the maps too may be deemed a manifestation of individual liberty. But, paradoxically, the same maps may also be viewed as tangible instruments of ruling powers to express the expansion of boundaries, taking cognizance of various physical features. Little wonder then that the Ptolemaic maps were so popular in the Renaissance. To the humanists, they imparted specific knowledge of the world beyond their own domains, even extending to the legendary cities of the east. To statesmen and *condottieri* like Lorenzo de' Medici and Federigo da Montefeltro, they held the potential for future conquest.

Our examination of the Ptolemaic city maps has revealed that they are still rooted in the culture of the medieval world. Affinities with city ideograms, navigational charts, crusader lore, and moral exempla abound. However, one prominent feature of the town views clearly presents their transitional nature, namely, the preponderance of domed centrally planned buildings. The type is used as a convention for religious edifices in all the views. Recall that in the Middle Ages, homage was given to the circle as the ideal form, visible in representations of Eden, the city of Jerusalem, or Dante's *Paradiso.* Whereas architects were supposedly striving to produce this centrally planned church emblematic of Renaissance principles, it is apparently but a commonplace in the workshops of the illuminators. The appeal of this building type to the quattrocento architect is equivalent to the search for a unified concept of space, the pure geometric form reflecting the absolute harmonic structure of the universe. Equated with the perfection of God, this centralized form could only be understood through mathematical symbols rooted in Neoplatonism, and, more fundamentally, in Euclidean geometry.[92] However, it is only the map of Milan that approaches a true circle, a fact commensurate with Bonvesin's panegyric to that city.

Correspondences with the medieval mind are reinforced by the hypothesis that the Ptolemaic maps may have been designed to promote a crusade. This is given support by the provenance and putative earlier date of the Paris codex and its patron King Alfonso of Naples. Urged on by a mission to free Christian Europe from the Turkish dominion, might he have ordered the addition of the cities to the modern maps? May we further posit that Alfonso's strong ties to Islamic culture led to the selection of the cities herein included? Alfonso's mission was allied to that of Pope Pius II, who, shortly before his death in 1464, had issued a bull proclaiming a crusade.[93] Consider too the pope's geographical studies. In view of the date of the earliest known city map of Vienna and the pope's detailed topographical description written during his sojourn there in 1458, might not Pius II be a force behind the creation of the Ptolemaic maps? Even the inclusion of Volterra in *Urb. lat. 277* finds earlier resonances in the passages in the *Commentaries* referring to "alum," the mineral used as a dyer's chemical that motivated the siege of the city.[94] Thus, mindful of the most significant political event of the quattrocentro, the fall of Constantinople, might the Ptolemaic views be among the last of the crusader guides to the Holy Land? Here are updated pilgrimage maps, the western cities represented by the principal Italian towns: Venice, port of call; Rome, bastion of the church; the eastern cities on the main trade route; the monuments depicted largely Christian. Above all, these maps may best be studied in conjunction with the literary accounts of pilgrims after the Crusades.[95]

An addition to the Urbino codex, it is the map of Volterra that constitutes a significant advance in mapmaking, anticipating the application of new artistic techniques to cartography. Not only may this view of Volterra be considered alongside Leonardo's contemporary drawing of the landscape of the Val d'Arno, dated 1473, but it looks forward to his strategic survey maps of Tuscan territories made for Cesare Borgia in the early 1500s. In his global view of Imola in 1502, Leonardo continued the cartographic revolution wrought by Alberti's circular map of Rome. Both revert to that series of circles which converge on the Palazzo Vecchio in Bruni's description of Florence, the city whose superiority and position at the center of a wide Tuscan landscape "is enclosed in a still larger orbit and circle."[96]

Although changing spatial and stylistic modes led to such pictorial representations as the perspective panels in Urbino, Baltimore, and Berlin, the medieval concept of the ideal city survived. In the mid-fifteenth century, Rome, the paradigmatic pagan and Christian city, replaced Jerusalem as the model City of God. Predictably, Florence in its turn would rival Rome. André Chastel has demonstrated that this view of Florence as the new humanistic Athens, the prototype of the ideal city, appears as the *civitas sapientiae* in a manuscript, ca. 1470, of Saint Augustine's *De Civitate Dei*.[97]

"Vedi nostra città quant' ella gira."[98] Dante's vision is perpetuated in some of the Ptolemaic maps and in the utopian cities of Renaissance theorists; it reaches its apogee in the maps of Alberti and Leonardo. Like all ideal concepts, this quest for a perfectly ordered geometric world mirroring the harmony of the heavens is at odds with the realities of contemporary Renaissance society. Reflecting the expanded boundaries of the quattrocento, the application of the new learning, the glorification of the city as harbinger of the new culture, and the revival of the ancient city republics, the Ptolemaic maps represent that moment heralding the birth of a new worldview—though one still marked by old ways of seeing and thinking and still subject to the ecclesiastical domain. Bearing the imprint of their medieval forebears, the "moralizing" maps for pilgrims, portolan charts for navigators, guides for Italian merchants—these maps celebrate the great metropolises east and west, an-

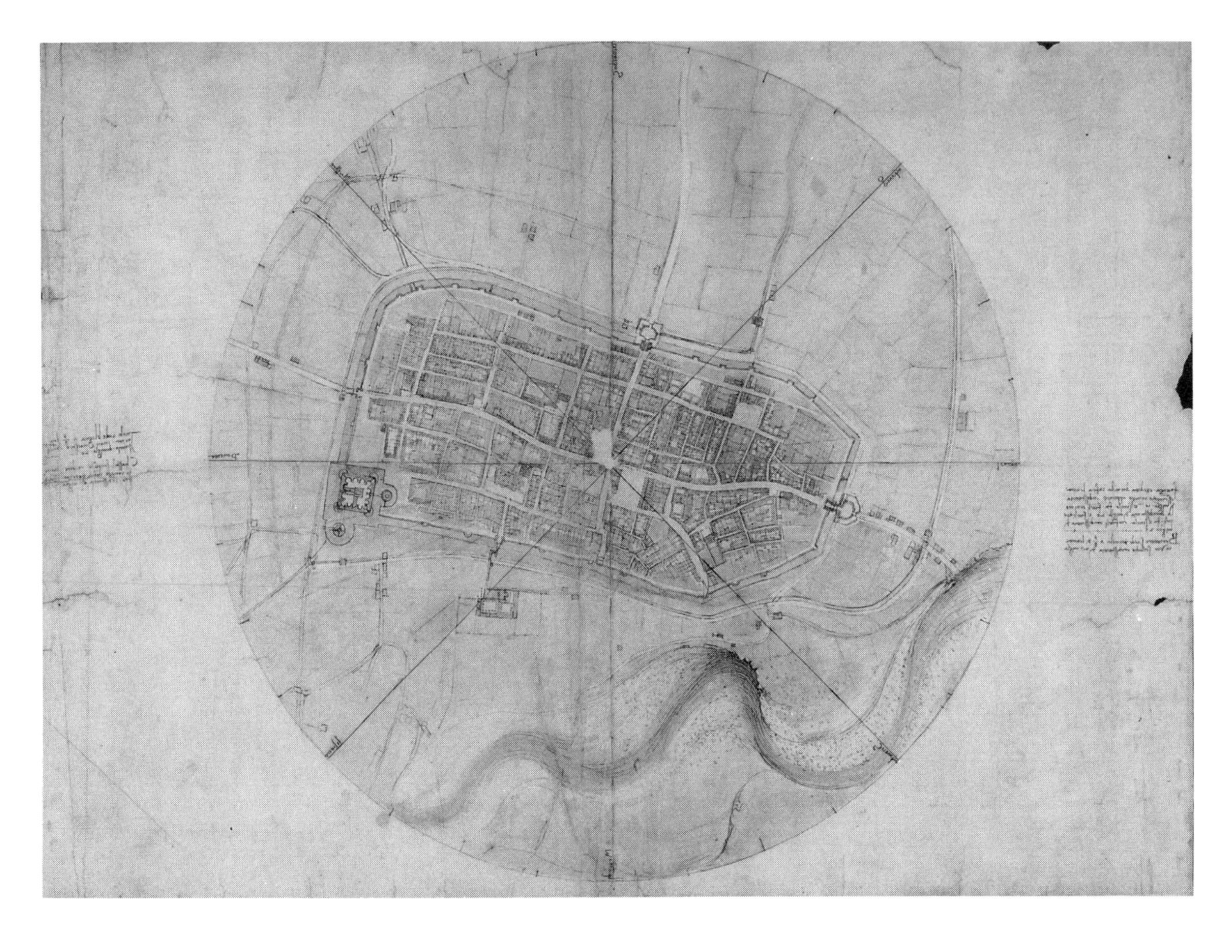

FIGURE 2.27. Leonardo da Vinci. Plan of Imola, 1502, Windsor, Cod. Atlantico 12284, pen and water color (photo: Reproduced by Gracious Permission of Her Majesty Queen Elizabeth II, Windsor Castle, Royal Library)

cient and modern, and further evoke the memory of Jerusalem, record the rebuilding of Rome, and sing the praises of Florence.

NOTES

1. Although the original manuscript of Jacopo d'Angelo's translation has not survived, manuscripts later in the century are clearly derived from *Vat. Urb. gr. 82.* Jacopo d'Angelo himself, secretary to the Papal Curia from 1401, inherited his task to translate Ptolemy's *Geography* from Emanuel Chrysoloras, who in turn had been persuaded to undertake this work by Paolo Strozzi. Despite the severe criticism of d'Angelo's translation, it was widely disseminated from the third decade of the fifteenth century. R. A. Skelton (*Maps: A Historical Survey of Their Study and Collecting* [Chicago and London, 1975], 38–39) notes forty-one manuscripts with the d'Angelo translation in the fifteenth century, made in Italy. Lloyd A. Brown (*The Story of Maps* [Boston, 1949], 154) hypothesizes that it was probably d'Angelo who changed the title from *Geographia* to *Cosmographia* (the title of the three manuscripts referred to in this study). For the sake of clarity, Ptolemy's opus will be called the *Geography* throughout this work.

See too Charles E. Armstrong, "Copies of Ptolemy's *Geography* in American Libraries" (*Bulletin of the New York Public Library* 62 [1962]: 105–13, esp. 107–11), where fifty-one printed editions are noted between 1475 and 1730. See Roberto Weiss, "Jacopo Angeli da Scarperia," *Medievo e Rinascimento: Studi in onore di Bruno Nardi* 2 (Florence, 1955), 811 n. 50, 812, 824.

2. The most complete discussion of the manuscripts as the work of Pietro del Massaio is in Joseph Fischer, *Claudii Ptolemaei, Geographiae, Codex Urbinas Graecus 82,* 3 vols. (London, 1932), 2:365–75. Essential data is also provided by Cosimus Stornajolo, *Biblioteca Apostolicae Vaticanae: Codices Urbinates Latini,* 3 vols. (1902), 1:253–54. Also, for Piero del Massaio (1420–ca. 1473–80), see Mirella Levi d'Ancona, *Miniatura e miniatori a Firenze dal XIV al XVI secolo* (Florence, 1962), 220–23; D. E. Colnaghi, *A Dictionary of Florentine Painters from the Thirteenth to the Seventeenth Centuries* (London, 1928), 56.

For Cominelli, see John W. Bradley, *Dictionary of Miniaturists, Illuminators, Calligraphers, and Copyists,* 3 vols. (London, 1988), 1:244–46. See too Albina de la Mare, "New Research on Humanistic Scribes in Florence," in *Miniatura fiorentina del rinascimento: 1440–1525,* ed. Annarosa Garzelli, vol. 1, Inventori e Catalogi Toscani (Florence, 1985), 18, 450–51, and Appendix, 505–6, for a list of Massaio's works. See too Berthold L. Ullmann, *The Origin and Development of Humanistic Script* (Rome, 1960), 21–57; David Woodward, "The Manuscript, Engraved, and Typographic Tradition of Map Lettering," in *Art and Cartography,* ed. David Woodward (Chicago and London, 1987), 174–212, esp. 179–80; also Otto Pacht, "Notes and Observations on the Origin of Humanistic Book Decoration," in *Fritz Saxl, 1890–1948,* ed. D. J. Gordon (London, 1957).

3. Skelton, *Maps,* 10; Brown, *Maps,* 36.

4. See facsimile edition (réduite), Paris, Bibliothèque Nationale. Département des manuscrits, *Géographie de Ptolemée* (Paris, 1926), introduction by H. Omont. See *A la découverte de la terre: dix siècles de cartographie* (Paris, 1979), 11–12, which dates this most beautiful of six manuscripts from the Latin translation by Jacopo d'Angelo toward 1470. See too De la Mare (*Miniatura* 1:210, 489), who notes that a copy of Ptolemy's *Cosmography*—"cholla pittura bellissima"—was sent to Naples in 1458. Cf. Tammaro de Marinis, *La Biblioteca napoletana dei rei d'Aragon,* ed. Ulrich Hoepli, 4 vols. (Milan, 1952), 2:141, which cites the acquisition of a "Cosmografia Tolomei" in 1456.

5. See Edward Luther Stevenson, *Geography of Claudius Ptolemy,* intro. by Joseph Fischer (New York, 1932). This translation is based on Greek and Latin manuscripts and important late fifteenth- and sixteenth-century printed editions. However, it should be noted that the Stevenson translation is deemed inadequate. The standard complete text remains that by C. F. A. Nobbe (1643–45), whereas the critical edition of Karl Müller (1883–1901) is incomplete.

6. See the life of Federigo da Montefeltro in Vespasiano da Bisticci, *Le Vite di uomini illustri,* ed. Aulo Greco, 2 vols. (Florence, 1970), 1:396. *Urb. gr. 82* was presented to the Florentine humanist Palla Strozzi. From the latter's heirs it went to Urbino in 1462, as recorded in Palla's last testament. See Aubrey Diller, "The Greek Codices of Palla Strozzi and Guarino Veronese," *Journal of the Warburg and Courtauld Institutes* 24 (1961): 313–17, esp. 316.

7. See Brown, *Maps,* 73.

8. Roberto Almagià, *Monumenta Cartographia Vaticana,* vol. 1, *Planisferi, carte nautiche. . . ,* 4 vols. (Città del Vaticana, 1944), 100, pls. 44–51.

9. Ibid.

10. Roberto Almagià, "Una carta della Toscana della meta del secolo XV," *Rivista Geographica Italiana* 28 (1921): 9–17.

11. Population estimates for the cities here included are rough, but with the exception of Cairo, the western cities are far denser. According to Josiah C. Russell (*Late Ancient and Medieval Population* [Philadelphia, 1958], 127), in the late fourteenth century the figures are: Venice, ca. sixty-five thousand, Florence, ca. fifty-five thousand, Rome, ca. thirty thousand, and Milan in 1463, ca. forty thousand. Russell (*Medieval Regions and Their Cities* [Vermont, 1972], 201, 208–9) also notes ca. 1200: Damascus, fifteen thousand; Jerusalem, ten thousand; and in the late thirteenth century, Cairo, sixty thousand; Alexandria, thirty to thirty-five thousand. Other estimated figures will be cited when discussing individual cities.

12. See, e.g., Mark Monmonier, *How to Lie with Maps* (Chicago, 1991), 2, who writes: *"A single map is but one of an indefinitely large number of maps that might be produced for the same situation or from the same data."*

13. See discussions of Massaio's maps in Leo Bagrow, *History of Cartography,* rev. and ed. by R. A. Skelton (Cambridge, Mass., 1964), 79–81. G. R. Crone (*Maps and Their Makers* [Folkstone, 1978], 36) notes the particular "method of representing relief" on Massaio's maps: "The highlands are marked off from the lowlands and their surface filled in by solid colors; though this method seems to represent all mountains as plateaus, it is possible to see in the dividing line between upland and lowland and in this use of color the prototype of form lines and layer coloring."

14. Jürgen Schulz, "Jacopo de' Barbari's View of Venice: Map Making, City Views, and Moralized Geography before the Year 1500," *Art Bulletin* 60 (1978): 425–74, esp. 458 n. 14. See also, John Harvey, *Topographical Maps* (London, 1980), 74.

15. Jürgen Schulz ("The Printed Plans and Panoramic Views of Venice (1486–1797)," *Saggi e Memorie di Storia dell' Arte* [Florence, 1970], 17–18) notes the revival of bird's-eye views in late fifteenth-century Tuscan and Sienese frescoes. See also Schulz, "Jacopo de' Barbari's View of Venice," 459.

16. See Francesco Cessi, "Vincenzo Dotto architetto padovano del XVII secolo," *Padova e la sua provincia,* 10,8 (1964): 9–14; and Silvano Ghironi and Guiliana Mazza, *Padova—Piante e vedute (1449–1869)* (Padua, 1985), 21, 22, 23.

17. Lionello Puppi and Mario Universo, *Padova* (Rome and Bari, 1980), 90, figs. 97, 98, 99 left, cat. 269, no. 1.

18. Lionello Puppi, "Iconografia di Padova ai tempi del Cornaro," in *Alvise Cornaro e il suo tempo* (Padua, 1980), 233. Cf. Verona and Mantua in Puppi and Universo, *Padova,* 163.

19. Puppi and Universo, *Padova,* 90.

20. Ibid., 90, 91, figs. 99 right, 100, 101, cat. 269, no. 2 (from S. Orsato, *Historia di Padova* [Padua, 1678], 114). Cf. too the district map of "Verona and her territory," dated 1439–40, which shows the city and its surrounding territory, emphasizing topographical features, as well as emblems of both civic and religious authority. See Lionello Puppi, *Ritratto di Verona* (Verona, 1978), 590–91; pl. 13, fig. 1; also Harvey, *Topographical Maps,* 59, pl. 4.

Ferrara and its territory are depicted in one of the earliest city plans, dated after 1326, giving some indication of principal buildings and reliable spatial relationships. A more coherent plan showing buildings in a crude perspective is that by Bartolino di Novara, dated ca. 1385. See Bruno Zevi, *Ferrara di Biaggio Rosetti, la prima città moderna europea* (Turin, 1960), figs. 61–70.

21. Harvey, *Topographical Maps,* 80, fig. 41. See Ferdinand Opll, *Wien im Bild Historischer Karten: Die Entwicklung der Stadt bis in die Mitte des 19. Jahrhunderts* (Vienna, 1983), pl. 1. The author notes that the plan dates to the early 1420s, as the river course of the Alserbaches changed in 1443 (13). See too Dana Bennett Durand, *The Vienna-Klosterneuburg Map Corpus of the Fifteenth Century: A Study in the Transition from Medieval to Modern Science* (Leiden, 1957), 28–29, for the role of northern and central Europeans in Renaissance cartography.

22. See Marino Sanudo, *Secrets for True Crusaders to Help Them to Recover the Holy Land,* trans. Aubrey Stewart (London, 1896). This text was originally written in 1312.

23. A. Manetti, *Vita di Filippo di Ser Brunellesco,* trans. H. Saalman (College Park, Pa., 1970). Also, J. B. Harley and David Woodward (*The History of Cartography* [Chicago and London, 1987], 1:497) note that Brunelleschi's technique was rarely followed, an exception being Nicolas of Cusa (d. 1464). See Harley and Woodward, *History of Cartography,* 1:488, 491–92, also for the use of a grid in the plan projected in 1306 for the city of Talamone, a coastal site acquired by Siena.

24. See Schulz, "Jacopo de' Barbari's View of Venice," 446–48.

25. As Ptolemy refers to seamen's charts, a relationship has been postulated between the ancient *periplus* (a narrative and a voyage around a coast) and the portolan (dating from ca. the thirteenth century). See Edward L. Stevenson, *Portolan Charts* (New York, 1911), 2–3.

A. E. Nordenskiöld ("Dei disegni marginale negli antici manoscritti del Dati," *La Bibliophilio* 3 [May–June 1901]: 49–55, esp. 55) has noted the connection between medieval portolans and the marginal illustrations of *La sphera,* and hypothesizes that Dati's work is perhaps a skillful compilation, tied to a Greco-Arabic original. Nordenskiöld finds it particularly revealing that these maps of the Mediterranean and Black Seas—the north coast of Africa, Syria and Asia Minor, Rhodes—chart a world that, prior to the Crusades, was under Moslem rule (53). In fact, the architecture of the towns and castles in these drawings is dominated by oriental and Islamic features.

26. As expected, most of the extant manuscripts are in Florentine libraries. The only printed version of the poem is that in the edition by Galletti (Florence, 1859, rpt. 1863), based on the 1514 Florence edition.

27. Harley and Woodward (*History of Cartography,* 490) also discuss examples of diagrams used as evidence in legal cases; e.g., problems concerning property rights in the use of waterways are cited from Bartolo da Sassoferrato, *De fluminibus seum tiberiadis,* 1355.

28. Brown, *Maps,* 96–97. Also Kenneth Nebenzahl, *Maps of the Bible Lands* (London, 1986), 9, fig. 3, illustrates a T-O map ca. 620 (the truncated cross within a disc) and cites its appearance in Isidore of Seville's encyclopedia (ca. 599–636).

29. See John Pinto, "Origins and Development of the Ichnographic City Plan," *Journal of the Society of Architectural Historians,* 35,1 (March 1976): 35–50. See also Harvey, *Topographical Maps,* figs. 95, 96.

30. Giovanni Fanelli, *Firenze* (Rome and Bari, 1980), 267, and n. 4; Giuseppe Boffito and Attilio Moro, *Piante e vedute di Firenze* (Florence, 1926), 7. See also Chiara Frugoni, *A*

Distant City: Images of Urban Experience in the Medieval World, trans. William McChaig (Princeton, N.J., 1991), 76–77.

31. Frugoni, *Distant City*, 79–80, cites Giovanni Villani, *Cronica*, Ed. A. Racheli (Trieste, 1857), VIII, 3, 171, who notes more than one hundred fifty towers belonging to citizens in Florence, and continues: "from far in the distance, she appeared the most beautiful and flourishing city for her limited area that could be found. And at this period she had a large population and was full of palaces and blocks of dwellings, and many people, for those times."

32. See Puppi and Universo, *Padova*, 82, fig. 96, cat. 269. Cf. Schulz, "Jacopo de' Barbari's View of Venice," 462–63.

33. John White, *Birth and Rebirth of Pictorial Space* (Cambridge, Mass., 1957), 94–95.

34. See Joan Gadol, *Leon Battista Alberti* (Chicago and London, 1969), 157.

35. Once dated ca. 1433 (see Richard Krautheimer, *Lorenzo Ghiberti* [Princeton, N.J., 1956], 315–16), a later date in the 1440s is now preferred. As Otto Lehmann-Brockhaus ("Alberti's 'Descriptio Urbis Romae,'" *Kunstchronik* 13 [1960]: 345–48) points out, the church of San Onofrio on the Gianicolo did not exist in the 1430s. Luigi Vagnetti ("Lo Studio di Roma negli scrittori Albertiana," *Problemi Attuali di Scienza e di Cultura* 209 [Rome, 1974], 97 n. 65, 109) states that the work was composed in 1452–53, during Alberti's second sojourn in Rome. Cecil Grayson (intro. to *Leon Battista Alberti: On Painting and On Sculpture*, ed. and trans. Cecil Grayson [London, 1972], 18–19) discusses problems in dating, but notes Alberti's concern with methods and instruments of measurement in these years when the later treatises were written.

36. George H. T. Kimble, *Geography in the Middle Ages* (London, 1938), 215. Bagrow and Skelton (*Cartography*, 35) state: "The geographical coordinates of these towns are given not in degrees, but in time; longitude is expressed in hours and minutes corresponding to the distance from the meridian of Alexandria and latitude in terms of the longest day, in hours and minutes." See too Gadol, *Alberti*, 180–92; she writes of Alberti's map of Rome: "like Ptolemy's maps [it] was a geometric picture of an empirically measured ground" (73).

37. See Gadol, *Alberti*, 167–80. She notes that Alberti found "*bearings* from his vantage point by means of an astrolabe-Horizon; and in this first geographical work on Rome, he found the *distances* between the points on 'course lines' as mariners did" (175).

38. Alberti, *On Sculpture*, discussing *dimensio* 124–29. See also Gadol, *Alberti*, 77–78.

39. Achille Ratti, "Due piante iconografiche di Milano del secola XV," *Atti del IV Congresso Geografico Italiano* (Milan, 1902), 603–15.

40. See Brown, *Maps*, 122–26. For maps of Milan, see Lucio Gambi and Maria Cristina Gozzoli, *Milano. Construzione di una città* (Bari, 1982).

41. Antonio Averlino detto il Filarete, *Trattato di architectura*, ed. Anna Maria Finoli and Liliana Grassi, 2 vols. (Milan, 1966), 1:53–63, 164–65, fols. 11v–14v. See 2:575 for Filarete's reference to a type of astrolabe used by Ptolemy.

42. See Leonardo da Vinci, Milan, Biblioteca Ambrosiana, Cod. Atl., fols. 199r and v. See also Carlo Pedretti, "Leonardo's Plans for the Enlargement of the City of Milan," *Raccolta Vinciana* (Milan, 1962), fasc. 19, 137–47, esp. 140–44; *Leonardo e le vie d'acqua* (Florence, 1983) fig. 3; Carlo Pedretti, *Leonardo architetto* (Milan, 1978), 57–64, regarding Leonardo's plan for the expansion of Milan in 1493. Ludwig Heydenreich (*Leonardo da Vinci*, 2 vols. [New York, 1954], 1:87–88) states that in the drawing at the bottom of the page Leonardo was "trying to convert the medieval planimetric scheme into an accurate picture with a central perspective . . . [thereby pointing] the way to the attainment of a fixed, yet entirely abstract viewpoint above the city, a point which at that time could only be represented through theoretical calculation and never attained in reality."

43. Touring Club Italiano, *Venezia* (Milan, 1985), 589–92.

44. See, e.g., Rodolfo Gallo, "A fifteenth-century military map of the Venetian territory of *Terrafirma*," *Imago Mundi* (Stockholm and Leiden, 1955), 55–57. He also notes the rarity of maps of the entire Veneto region before the mid-sixteenth century.

45. Giorgio Bellavitis and Giandomenico Romanelli, *Venezia* (Bari, 1985), 53–60. The plan is in the Bibl. Naz. Marciana, Cod. Lat. Z = 399, no. 1610, f. 7n.

46. Schulz, "Printed Plans," 16 and n. 20; Schulz, "Jacobo de' Barbari's View of Venice," 440–41 and 445 n. 60, where he cites the original plan made in the early twelfth century by a surveyor from Milan (see fig. 9).

47. Harley and Woodward, *History of Cartography*, 478.

48. Roberto Almagià, *Monumenta italiae cartographiae* (Florence, 1929), 11–12.

49. Giovanni Fanelli, *Firenze* (Rome and Bari, 1980), 53, citing *Epistola, ecc.*, ed. Lapo da Castiglionchio, (Bologna, 1753), 47. See also Harvey, *Topographical Maps*, 78.

50. Hans Baron, *The Crisis of the Early Italian Renaissance*, 2nd ed. (Princeton, N.J., 1966), 200–201.

51. See Mario Baratta, *Piante di città*, tav. 17. Kenneth Clark (*A Catalogue of the Drawings of Leonardo da Vinci . . . Wind-*

sor Castle [Cambridge, 1935], 142) cites Richter, § 1014 (recto), listing eleven city gates. See Jean Paul Richter, *The Literary Works of Leonardo da Vinci,* 2 vols. (London, 1939), 2:182, § 1006, regarding notes on the course of the Arno, written alongside a "sketch for a completer map."

52. See Torgil Magnuson, "Studies in Roman Quattrocento Architecture," *Figura* (Stockholm, 1958), 115 ff. See also Amato Pietro Frutaz, *Il Torrione di Niccolo V in Vaticana* (Città del Vaticana, 1956), 26 n. 4; and Ludwig Heydenreich and Wolfgang Lotz, *Architecture in Italy, 1400–1600* (Harmondsworth, 1974), 51–52.

53. Giustina Scaglia, "The Origins of an Archaeological Plan of Rome by Alessandro Strozzi," *Journal of the Warburg and Courtauld Institutes* 27 (1964): 137–63.

54. See A. P. Frutaz, *Le Piante di Roma,* 3 vols. (Rome, 1962, For the Strozzi map, see 1:89, pp. 140–42; 2: pl. 159. For Massaio's maps, see 1:87–88, 90, pp 137–40, 42–44; 2: pls. 157–58, 160. Further support for an archetype dating before 1453–56 is the absence of the angel on the Castel San Angelo, only completed in 1453. See too Roberto Weiss, *The Renaissance Discovery of Classical Antiquity* (Oxford, 1973), 92 n. 2.

55. Brockhaus, "Descriptio," 346. Leon Battista Alberti, *L'Architettura* (*De Re Aedificatoria*), trans. G. Orlandi, 2 vols. (Milan, 1966), 2: bk. 6, chap. 1, 440, expresses his concern with the restoration of antiquities—another motive for his archaeological reconstruction of Rome.

56. From H. Vast, *Le Cardinal Bessarion* (Paris, 1878), 212 (my trans. from the French); Latin text, 454–56.

57. Mark Girouard (*Cities and People* [New Haven, Conn., 1985], 4) notes 12 miles of wall, 37 gates, and 486 towers. See also Wolfgang Müller-Wiener, *Bildlexikon zur Topographie Istanbuls* (Tübingen, 1977).

58. Girouard, *Cities,* 10–11. See also Eve Borsook, "The Travels of Bernardo Michelozzi and Bonsignore Bonsignore in the Levant (1497–98)," *Journal of the Warburg and Courtauld Institutes* 36 (1973): 145–97, esp. 158–59. See too Franz Babinger, *Mehmed the Conquerer and His Time,* trans, W. C. Hickman (Princeton, 1978).

59. Richard Krautheimer, *Three Christian Capitals: Topography and Politics* (Berkeley and Los Angeles, 1983), 49.

60. Christophe Buondelmonti, *Description des iles de l'Archipel, Version grecque . . . avec une traduction française et un commentaire* (Paris, 1897), 243–44. Is this the Column of Constantine, with Constantine as Helios, the New Sun, lance in left hand, glove in right, as depicted in the thirteenth-century copy of a second-century map, the *Tabula Peutingeriana?* See Krautheimer, *Capitals,* 56; also 62–63. See also Roberto Weiss, "Christoforo Buondelmonti," *Dizionario Biografico degli Italiani* (Rome, 1972), 15: 198–200. For Buodelmonti's views of Constantinople, see Giuseppe Gerola, "Le vedute di Constantinople di Christoforo Buondelmonti," *Studi Bizantini e Neohellenici* (Rome, 1931), 247–79. Ian R. Manners, "Constructing the Image of a City: The Representation of Constantinople in Christopher Buondelmonti's *Liber Insularum Archipelagi,*" *Annals of the Association of American Geographers,* 87:1 (1997): 72–102, focuses on copies of Buondelmonti's original, following the fall of the city; these maps stress its Christian heritage, thereby ignoring the Ottoman monuments constructed after 1453.

61. Anna C. Esmeijer ("Hierusalem Urbs Quadrata," *Divina Quaternitas* [Amsterdam, 1978], 73–91) discusses a "visual exegetic scheme" with particular attention to Heavenly Jerusalem, the ideal city with a circular ground plan divided orthogonally into four (73, 156, notes).

62. Hilda F. M. Prescott, *Friar Felix at Large: A Fifteenth-Century Pilgrimage to the Holy Land* (New Haven, Conn., 1950), 124–25.

63. See Nebenzahl, *Maps,* 8–13.

64. See Felix Fabri, *The Wanderings of Felix Fabri* (London, 1897), 8:534–38.

65. According to Zev Vilnay (*The Holy Land in Old Prints and Maps* [Jerusalem, 1965], 51), Hakeldama is a corruption of Aramaic. Niccolo da Poggibonsi (*A Voyage beyond the Seas (1346–50),* trans. T. Bellorini and E. Hoade [Jerusalem, 1945], 36) describes this field (cf. Matt. 27:2) as the Holy Field, a goal of pilgrims, for which "there is a very big indulgence." Frank E. Peters (*Jerusalem: The Holy City in the Eyes of Chroniclers, Visitors, Pilgrims, and Prophets from the Days of Abraham to the Beginnings of Modern Times* [Princeton, N.J., 1985], 457 and n. 20) states that this was the burial place for the poorer Christians, who died in the city. See too Theoderich (*Guide to the Holy Land,* 1896, trans. Aubrey Stewart, 2nd ed. [New York, 1986], 7), who notes "In the field of Acheldemach, which is only separated from the [Jehosophat] valley, is the pilgrims' burial ground, in which stands the Church of St. Mary." For biblical sources of Aceldama ("Field of Blood"), see Jerome Murphy-O'Connor, *The Holy Land* (Oxford, 1980), 76–77.

66. See Peters, *Jerusalem,* 479–80. He notes the difficulty of determining demographic figures, but estimated population is usually cited as about twenty to thirty thousand (408). (Meshullam of Volterra, ca. 1480, numbers ten thousand Muslim households and two hundred fifty Jewish ones.)

67. Cf. Karl Baedecker (*Palestine and Syria* [Leipzig, 1898], 76), who gives the dates of 1160–89 for the Ro-

manesque tower built by the crusaders and the octagonal-domed chapel of the forty martyrs.

68. Theoderich (*Guide,* 22) describes the beauty of the Church and Hospital of Saint John the Baptist and the excellence of its facilities. See too H.F.M. Prescott, *Jerusalem Journey: pilgrimage to the Holy Land in the fifteenth century* (London, 1954) 118, 129, passim.

69. John Wilkinson (*Jerusalem Pilgrims before the Crusades* [Jerusalem, 1977], 10) dates Arculf's pilgrimage between A.D. 679 and 688.

70. Maps of Damascus, including reconstructions of the ancient city, are found in guidebooks and in pilgrimage literature. See John Murray, *Handbook for Travellers in Syria and Palestine* (London, 1903), 308–20; Baedecker, *Palestine and Syria,* 340–43. See too Edward Robinson, *Biblical Researches in Palestine and the Adjacent Regions,* 3 vols. (London, 1856), 3:446–49.

71. See J. L. Porter, *Five Years in Damascus,* 2 vols. (London, 1855), 1:63. See too Robinson, *Biblical Researches,* 3:462; and Christine P. Grant, *The Syrian Desert* (London, 1937), 90–91 and maps.

72. See map in Hinru Riad, *Alexandrie: Guide archéologique . . .* (Alexandria, [19??]); and Peter M. Fraser, *Ptolemaic Alexandria,* 3 vols. (Oxford, 1972), 1:271; 2:33 n. 81.

73. Fraser, *Ptolemaic Alexandria,* 1:29; 2:210–12. Cf. Strabo's description: "It has the shape of a fir-cone, resembles a rocky hill, and is ascended by a spiral road; and from the summit one can see the whole city lying below it on all sides" (*Geography,* trans. H. J. Jones, 8 vols. [Cambridge, Mass., 1949], 8:17.1.10, p. 41.

74. Fraser, *Ptolemaic Alexandria,* 1:238–39; see also W. H. McLean, *City of Alexandria Town Planning Scheme* (Cairo, 1921), 13.

75. Strabo, *Geography,* 8:17.1.6, p. 29: the Greeks "gave them as a place of abode Rhacotis . . . which is now a part of the city of the Alexandrians which lies above the shiphouses, but was at that time a village."

76. E. M. Forster, *Alexandria* (New York, 1961), 51.

77. Fraser, *Ptolemaic Alexandria,* 2:19–20, 43–46. See also André Bernand, *Alexandrie la Grande* (Paris, 1966), 105–6.

78. In his voyage to Egypt in 1171–73, Benjamin of Toledo praised the "wide streets of the city" and the beauty of the buildings. See John K. Wright, *The Geographical Lore of the Time of the Crusaders* (1925; New York, 1965), 299.

79. Poggibonsi, *Voyage,* "On the Castle of the Sultan," 88. The account of Ludolph von Suchem, *Description of the Holy Land,* trans. Aubrey Stewart, 1895 (rpt. New York, 1971), 130, appears to transpose this citadel to Damascus.

80. For the complex history of Saladin's citadel, see K. A. C. Creswell, *Muslim Architecture in Egypt,* 2 vols. (Oxford, 1952–59), esp. vol. 1 and vol. 2, chap. 5. See also Dorothea Russell, *Medieval Cairo* (New York and Toronto, 1962), 195–210.

81. See R. J. Mitchell, *The Spring Voyage: The Jerusalem Pilgrimage in 1458* (New York, 1964), 153, who quotes Roberto da Sanseverino: "The chief streets are well furnished with all sorts of things, but not in the manner of ours, nor do they have such beautiful houses. . . . Theirs are all of stone, . . . their ceilings are of gold and blue and they have many carpets."

82. Oleg V. Volkoff, *Le Caire, 969–1969* (Cairo, 1970), 9.

83. Cf. The similar vernacular in Alexandria. Note too that open loggias exist in late medieval Florence, e.g., in the Palazzo Davanzati.

84. Georg Braun and Franz Hogenberg, *Civitates Orbis Terrarum,* 4 vols. (Cologne, 1572–98), 1:55. See also Stanley Lane-Poole (*The Story of Cairo* [1902; Nendeln-Liechtenstein, 1971], 96), who notes two nilometers on Roda; Russell, *Medieval Cairo,* 107–8; and Karl Baedecker, *Egypt* (Leipzig, 1929), 112–13.

85. Here we might recall Sanudo (*Secrets,* 60), who repeats an error common in pilgrims' accounts: "Five leagues from Babylon there are some triangular pyramids, exceeding lofty, which are said to have been Joseph's granaries."

86. My translation; cited in Enzo Carli, *Volterra nel medioevo e nel rinascimento* (Pisa, 1978), 12: "Volterra—sopra un gran monte, ch'è forte ed antica—quanto in Toscana alcun' altra terra."

87. A similar type of weapon is depicted in *Vat. Urb. lat. 282,* Giov. Angelo Terzone, *Delle cose spettanti all militia,* early sixteenth century, fols. 80r–82r; *De la bombarda* is shown against a mountainous terrain.

88. Bernardino Baldi, *Vita e fatti di Federigo di Montefeltro,* estratta da MS: inedito della Biblioteca Albani e corredati di Osservanzioni, 3 vols. (Rome, 1824), 3:212–27, esp. 217, 222–23.

89. Benedetto Dei, *La Cronica dall' anno 1400 all' anno 1500* (Florence, 1984), 74–76. Dei also takes this opportunity to list the provenance of the different troops of soldiers. See too Archivio Stato di Firenze May, letter 7 maggio 1472 (2.84, 223–24).

90. *Un Toscano del '400. Poggio Bracciolini, 1380–1459* (Terranova Bracciolini, 1980), 141. See James Dennistoun, *Memoirs of the Dukes of Urbino, 1440–1630,* 3 vols. (London, 1909), 1:213: Bracciolini, seeking a patron, thought it would be expedient to inscribe the manuscript to "one who had just

crowned their arms with signal success." Federigo's rapport with Florentine humanists is also seconded in Cristoforo Landini's dedication to him of the *Disputationes Camaldulenses,* written after the sack of 1472 (*Vat. Urb. lat. 508*). See Cristoforo Landini, *Scritti critici e teorici,* ed. Roberto Cardini, 2 vols. (Rome, 1974), 22, 66.

91. See Frugoni, *Distant City,* 54; Baron, *Crisis,* 191–211.

92. Rudolph Wittkower (*Architectural Principles in the Age of Humanism* [New York, 1965], 29) has encapsulated the significance of this form in the minds of Renaissance humanists: "For the man of the Renaissance this architecture with its strict geometry, the equipoise of its harmonic order, its formal serenity, and, above all, with the sphere of the dome, echoed and at the same time revealed the perfection, omnipotence, truth, and goodness of God."

93. *Memoirs of a Renaissance Pope: The Commentaries of Pius II,* an abridgment, trans. F. A. Gregg (New York, 1959), 236–39, 369.

94. *Ibid.,* 356. The pope's words confirm the preciousness of the material. Or consider how, in 1462, a wealthy businessman confronts Pius II: "Today I bring you victory over the Turk. Every year they wring from the Christian more than 300,000 ducats for the alum with which we dye wool various colors . . . though by God's grace a vein of alum has been discovered which continually puts us in greater debt to the Divine Mercy and encourages us to protect religion" (253). See Charles Singer, *The Earliest Chemical Industry: An Essay in the Historical Relations of Economics and Technology illustrated from the Alum Trade* (London, 1948), 140–41. Discussing the new mercantilist doctrine of Florence, Singer presents the city's declaration of war on Volterra in 1472, under the command of Federigo da Montefeltro, as a dramatic illustration of the trade's importance. Not only did the victors annex the city, but "the alum works were leased to the Florentine *Arte della lana* 'by reason of their urgent need of that substance.'"

95. See the chapter on "The Islamic City" in Enrico Guidoni, *La Città europea, formazione e significato del IV all' XI secolo* (Milan, 1978), 57–59 and notes.

96. Baron, *Crisis,* 204. See too Martin Kemp, *Leonardo da Vinci. The Marvellous Works of Nature and Man* (Cambridge, MA, 1981), 228: "The Imola map is the most magnificent surviving product of the Renaissance revolution in cartographic techniques," Kemp also discusses Leonardo's debt to Alberti, and further, the map's inclusion of compass bearings and distances of nearby towns (228–30).

97. André Chastel, "Un episode de la symbolique urbaine au XVe siécle: Florence et Rome, Cités de Dieu," *Urbanisme et architecture: Etudes écrites et publiées en l'honneur de Pierre Lavedan* (Paris, 1954), 75–79, esp. 78. The manuscript is in the New York Public Library, Spencer Collection.

98. "See our city, how wide is its circuit!" (trans. C. Singleton *The Divine Comedy* (Princeton, 1975) Dante, *Paradiso* 30:130; cited in Chastel, "Un episode de la symbolique," 76.

THREE

Urbs and *Civitas* in Sixteenth- and Seventeenth-Century Spain

Richard L. Kagan

Writing in the seventh century, Isidore of Seville, the famous encyclopedist and theologian, defined the term city in the following way: "A city [*civitas*] is a number of men joined by a social bond. It takes its name from the citizens who dwell in it. As an *urbs,* it is only a walled structure, but inhabitants, not building stones, are referred to as a *city.*"[1] Put simply, Isidore was attempting in this definition to distinguish between the city as community (*civitas*) and the city as an urban architectural entity (*urbs*).

As rudimentary as this definition may appear, it is important for understanding the various ways in which cities were conceived during subsequent periods, especially the Renaissance. The idea of the city as urbs can be found, for example, in the writings of numerous architectural theorists of this period, many of whom, L. B. Alberti included, argued that a city's nobility was inextricably linked to the layout of its squares and streets as well as the design and magnificence of its buildings. Toward the end of the sixteenth century, Giovanni Botero, in his essay, *On the Greatness of Cities* (1588), would make a similar argument, although this Italian thinker believed that a city's grandeur depended primarily on the number and importance of its inhabitants.[2]

This particular notion of city vied with another, espoused mainly by political theorists and theologians, who conceived of the city less in terms of architecture and demography than in those of community. Following Aristotle's definition of *polis,* these authors tended to define city as a *res publica,* a well-governed community or republic, and argued that a city's grandeur depended less on the quality of its buildings than the quality of its government.

In Spain, such ideas were central to the *Siete Partidas,* the famous thirteenth-century Castilian law code in which a town (*pueblo*) was doubly defined as "a place surrounded by walls" as well as the "communal gathering of men—the old, those of middling age, and the young."[3] They can also be found in the work of various theologians and political writers, starting with Fray García de Castrojérez and Francesc Eiximenis in the fourteenth century and later in the treatises of Rodrigo Sánchez de Arevalo (1404–70), Alonso de Castrillo (fl. 1521), and Diego Pérez de Mesa (1563–1616).[4] On a somewhat less theoretical plane, they are found in the work of Dámaso de Frias, a sixteenth-century lawyer from Valladolid, who defined city as "as a gathering of many families united in law and in custom for the purpose of providing all the things necessary for life." Frias, moreover, took issue with prevailing architectural theories when he wrote that the *grandeza* and *nobleza* of a particular city had little to do with the magnificence of its public buildings and squares. A city's nobility, he believed, depended primarily on the deeds and virtues of its inhabitants.[5] Frias, in short, expressed a notion of city that accorded precedence to people, not bricks, a position echoed in most sixteenth- and seventeenth-century histories of Spanish cities as well as in Sebastián de Covarrubias's dictionary of the Spanish language, first published in 1611. Borrowing almost directly from Isidore, his entry for *ciudad* read as follows:

> City. From the Latin noun *civitas* . . . A city is a collection of citizens who have congregated together in order to live in the same place under the same laws and government. City is sometimes understood as buildings; this corresponds to the Latin noun *urbs*. City can also refer to *regimiento* or *ayuntamiento* [terms equivalent to civic government] and, in the Cortes, to the agent [*procurador*] who represents a particular city.[6]

As Covarrubias's definition suggests, the notion of a city as a physical or topographical phenomenon—a space bounded by walls—existed but took second place to the idea of the city as community. By the seventeenth century, in fact, this particular conception of city achieved the virtual status of a paradigm in Spanish thought. It dominated and, in a sense, structured the way Spaniards wrote chronicles and histories of their native towns, and it directly influenced the way many artists and draftsmen depicted places to which they were closely connected. Cities tended consequently to be "mapped" according to criteria that often had little to do with "description," the term generally used to refer to the evocation of places, either in words or in images, with a modicum of topographical accuracy and verisimilitude.[7] Rather, the aim was to capture the very essence or soul of the city, the particular moral values believed to ennoble a city and to accord it a unique place in history, both human and divine.

With Isidore's twin definition of city serving as a point of departure, this essay will briefly examine various plans and views of sixteenth- and seventeenth-century Spanish cities. I am particularly interested in making a comparison between two distinct but sometimes overlapping conventions or modes of urban representation. One, to put it simply, is the "chorographic" view, the city as seen by individuals who attempted to offer, in so far as the technical capacities of the era allowed, a complete and comprehensible visual record of a particular place. Chorography, according to Ptolemy, attempted "to describe the smallest details of places" as well as "to paint a true likeness of the places it describes."[8] Petrus Apianus, the geographer who was one of the Emperor Charles V's teachers, defined chorography in the following way:

> Chorography is the same thing as topography, which one can define as the plan of a place that describes and considers its peculiarities in isolation, without consideration or comparison of its parts either among themselves or in relation to other places. But at the same time chorography carefully takes note of all particularities and properties, as small as they may be, that are worth noting in such places, such as ports, towns, villages, river courses, and all similar things, including buildings, houses, towers, walls, and the like. The aim of chorography is to depict a particular place, just as an artist paints an ear or an eye or other parts of a man's head.[9]

Chorographic images of a city, then, were those that tended toward completeness and precision. Most commonly, they were the work of professional cartographers, engineers, and surveyors, but chorographic views were also produced by artists commissioned to document a monarch's travels or to give visual expression to the landholdings and possessions of an individual nobleman. Although difficult to categorize, chorographic images of Spanish cities included both prospects and plans, many of which were expressly designed to be engraved or printed and distributed to a wide audience. Typical of such views were the drawings of various cities in France, Italy, and the Low Countries executed by the Flemish artist, Anton van den Wyngaerde, in the middle years of the sixteenth century. Van den Wyngaerde's Spanish townscapes will be discussed below.

The other mode of representation is what might be characterized as the "communal" or "communicentric" view of the city.[10] Although these views often shared some of the descriptive elements associated with chorography, they did not necessarily seek to convey, prima facie, a "true likeness" of a city. They tended rather in the direction of metaphor and sought to define, via the image of *urbs,* the meaning of *civitas:* the idea of the city as a human community or well-governed republic endowed with a character, history, customs, and traditions uniquely its own. Communicentric views, therefore, often had a didactic aim, typified in some respects by Ambrogio Lorenzetti's fourteenth-century allegory of *Good and Bad Government* in Siena's town hall, a fresco depicting an idealized city intended to serve as a political primer for the councillors of the Sienese commune.[11] Communicentric views were also frequently connected to local religious cults and practices along with the commemoration of historical events of local importance. It follows that communicentric views were intended, in the first instance at least, primarily for local consumption rather than for wide distribution. As such, these views seem to offer a clue to the various ways the inhabitants of a particular community visualized—and in so doing, defined—the city in which they lived. In this sense they may help us understand what Kevin Lynch has described as a city's "public image," a term he defines as the common mental pictures of the local cityscape carried by large numbers of a city's inhabitants.[12]

Underpinning this comparison between chorographic and communicentric views is my dissatisfaction with much of the existing literature on city views, particularly that which concerns the history of the genre's development in early modern Europe.[13] This history is traditionally presented as if it were both linear and progressive, moving steadily but relentlessly from crude to refined, artistic to scientific, portrait to plan, view to map. Generally speaking, this Whiggish schema is correct, but as this essay will suggest, it tends not only to oversimplify the multifaceted character of city views but also to ignore the extent to which their character was shaped by the purposes to which they were originally put.

Chorographic Views

A convenient starting point for a discussion of chorographic images of Spanish cities is the unusual bird's-eye view of the Old Castilian town of Aranda de Duero, dated 1503 (figure 3.1). Comparable in time if not in quality to Leonardo da Vinci's famous ichnographic view of Imola, the circumstances surrounding the commission of this watercolor—a lawsuit arising from plans to cut a new street through the town center—helps us to understand its contents.[14] In this instance, the lawsuit reached the king's councillors who, needing additional information about the circumstances of the case, sent a certain Francisco Gamarra on an investigative trip to Aranda. Gamarra then commissioned a cartographer—we do not know who—to prepare a plan to help the judges to reach an equitable decision. The resulting orthogonal view of Aranda—the first of its kind for any Spanish city—is not entirely accurate. Its rounded shape, for

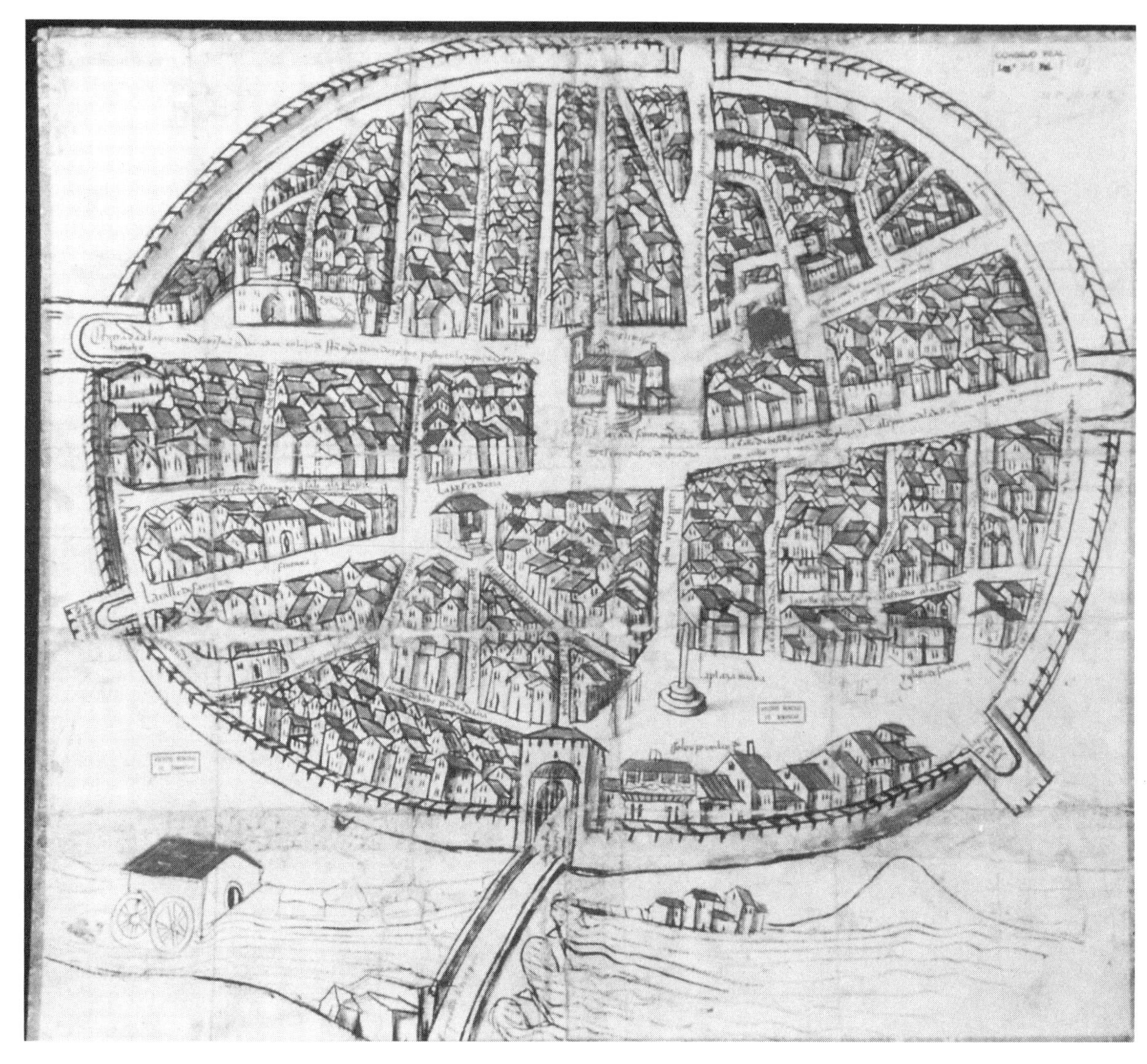

FIGURE 3.1. Anonymous. Aranda de Duero. Plan. 1503. Archivo General de Simancas. Mapas, Planos y Dibujos X-1.

example, suggests that Gamarra modeled his view of Aranda after some archetypical image of an idealized town. In its outline, then, the Aranda represented here is akin to a *typus* or conventional view of a town of a kind included in many medieval maps and portolans. As for the rest of Aranda, there is little reason to suppose that Gamarra altered or otherwise distorted the displacement of streets and buildings for pictorial effect or ideological purpose. Rather he endeavored to provide the judges with an unadorned but comprehensible plan of Aranda's streets and

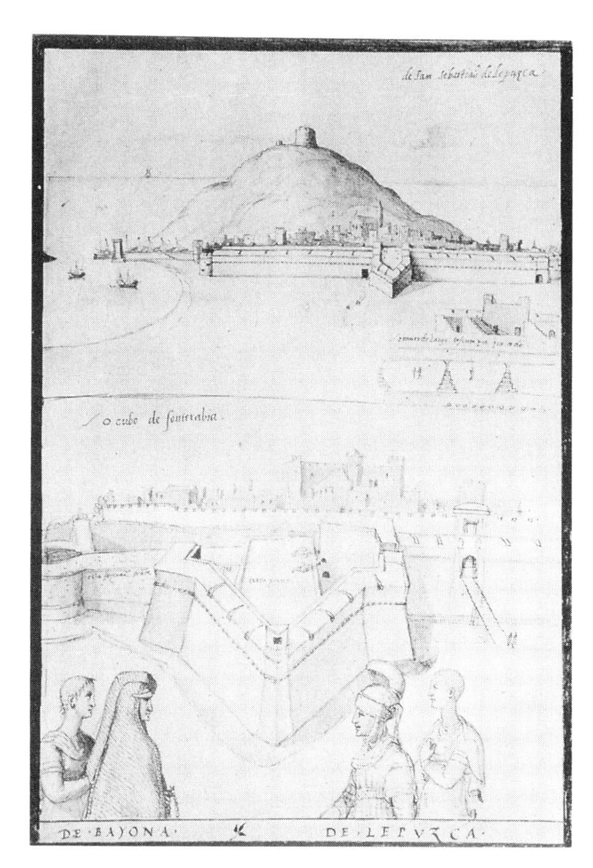

FIGURE 3.2. Francisco d'Olanda. Fuenterrabía and San Sebastián. C. 1540. Source: *Os desenhos das antigualhas.*

principal monuments. Despite its conventional qualities, therefore, the view constituted a description offering a detailed representation of Aranda as *urbs.*[15]

Description of a similar sort was also the purpose of Francisco de Holanda's perspective view of the port city of San Sebastián and the accompanying view of the Spanish fortress at Fuenterrabía in the 1540s (figure 3.2). This drawing, appended to Holanda's Italian sketchbook of city views, was executed to help his patron, King John III of Portugal, further his understanding of the latest European developments in fortification techniques.[16] Wherever he went, therefore, Holanda took pains to represent, as accurately as possible, encircling walls, bastions, and other defenses. In this instance, Holanda utilized perspective in order to reinforce the notion that his image of San Sebastián was indeed a "true likeness" of the city. Moreover, he utilized a low vantage point that allowed for the detailed depiction of the city's walls and other fortifications but which obscured the city's churches and other monuments, that is, the urban artifacts around which most sixteenth-century artists customarily constructed their views, often at the expense of topographical accuracy. In this instance, however, Holanda, instead of adhering closely to what the dictates of chorography demanded, included in the foreground of his drawing of the fortress of Fuenterrabía four figures, two of whom appear to be Romans while the others are dressed in Basque costume typical of the era. They suggest that Holanda, in addition to chorography, had other artistic and cultural concerns.

A different approach to the city as *urbs* appeared in the many woodcuts illustrating Pedro de Medina's *Libro de las grandezas y cosas memorables de España* (Seville, 1548), one of the most comprehensive geographical treatises of its day. Medina, a cosmographer attached to Seville's House of Trade (*Casa de Contratación*), wrote it to help the future Philip II learn about "the things of this, your Spain," and presented it to the prince as "a manual or aide-memoire of the most distinguished and important jewels—*'joyas'*—that Spain possesses."[17]

These "joyas" were Spain's cities, and the book itself was a compendium of approximately four hundred fifty city-biographies that incorporated information about geographical location, agriculture and commerce, principal monuments, and historical events of local importance.

FIGURE 3.3. Segovia and Valencia as illustrated in Pedro de Medina, *Libro de las grandezas* (1548)

Approximately one-fourth of the entries were accompanied by a woodcut, but the scope of the enterprise prevented Medina from offering a true visual likeness of the cities he described. For the most part, therefore, the illustrations included in the volume had little to do with the cities described in the text. Rather these cities, as in so many other early sixteenth-century atlases and geographical books, were represented by means of a *typus* meant to convey the general concept of *urbs* and, along with it, the sense of the city as an independent, self-governing entity.

This lack of specificity, so different from the previous examples of Aranda de Duero and San Sebastián, explains why the views in Medina's book were interchangeable. In the book's first edition, two different images represented all of Spain's towns. The first, employed mainly for smaller towns, depicted a building surrounded by walls, an image that evoked the traditional understanding of a town as a walled enclave set apart from the surrounding countryside.[18] The second, reserved for larger, more important cities, incorporated a fortress evidently intended to serve as a symbol of governmental authority. In addition, views illustrating maritime cities such as Valencia included a reference to the sea, whereas those for interior cities included references to mountains or hills (figure 3.3). Otherwise, apart from a label indicating the name of the city in question, these images made no effort to represent a specific place. Within this repetitive and somewhat monotonous constellation only three cities—Granada, Seville, and Toledo—received anything like specific treatment; the others offered little more than a topos of urban life.[19]

Far livelier were the cityscapes of Georg Hoefnagel, the Flemish artist whose views of Spanish and other European cities circulated widely throughout Europe in Georg Braun and Franz Hogenburg's *Civitates Orbis Terrarum,* the first volume of which—three were eventually published—appeared in 1572. Hoefnagel was an excellent draftsman; artistically, his townscapes were among the finest produced in sixteenth-century Europe. He also took pride in the fact that he rendered his views *ad vivum,* a designation meant to suggest they were based on first-hand observation and therefore topographically correct.[20]

Despite these boasts few of Hoefnagel's Spanish views—all dating from the 1560s—offered a "true likeness" of the towns they purported to represented. They were not "descriptions," in the Ptolemaic sense, or even "portraits" in the sense of faithful representations of what Hoefnagel had seen, but rather "constructions" designed to convey the impression of exactitude and precision.[21] Hoefnagel also adopted a compositional scheme that generally subordinated the representation of *urbs* to the particularities of daily life. Accordingly, he generally relegated the city to the background of his compositions, where it appeared more often than not as an indistinguishable blur of rooftops, interrupted only occasionally by a church tower or some other identifiable (and accurately rendered) monument. In contrast, the foreground featured fanciful genre scenes illustrating some of the more unusual features of Spanish life. What was pictured here varied, but on the whole these genre scenes reflected Hoefnagel's attempt to make his views interesting for the readers of the *Civitates*, most of whom, presumably, lived in northern Europe.[22] These scenes were especially prominent in his views of Andulusia, in southern Spain. There he had an eye for typically Mediterranean commercial activities such as the *almadraba* or tuna fishery near Cadiz, or the impressively large storage jars (*tinajas*) in which the peasants of Antequera stored olive oil and wine—two staples of Mediterranean life. He also had a particular interest in the *moriscos*, remnants of Spain's Muslim population who, especially for an artist from northern Europe, must have been both strange and exotic. Hoefnagel was also on the lookout for unusual, peculiarly Spanish social customs, notably the *juego de cañas*, a type of armed joust, that dominated the foreground of his view of Jérez de la Frontera. Similarly, his view of Seville placed the city in the distant background in order to accommodate a foreground that focused on the public shaming (*vergüenza*) of a horned and cuckolded husband being paraded on top of a burro (figure 3.4).

According to Lucia Nuti, these genre scenes reflected Hoefnagel's "merchant's eye"—he came from a family of Antwerp traders—and belonged to a tradition of fanciful traveler accounts that began with Marco Polo.[23] More specifically, Hoefnagel's interest in Spanish folklore owed much to Christoph Weiditz, the Augsburg engraver whose *Trachtenbuch* (Costume book), compiled during the course of a visit to Spain begun in 1529, recorded both the costumes and customs of the Spanish peasantry.[24] Hoefnagel's interest in the social and cultural rather than strictly urban aspects of Spanish life also belonged to sixteenth-century Europe's "curiosity movement," a concern for any unusual social and natural phenomena. Occasioned in large part by the "discovery" of America, the movement was subsequently demonstrated in the formation of *cabinets des curiosités* as well as in Hoefnagel's own fascination with the diversity of species, bugs and insects included, inhabiting the natural world.[25] But whatever the precise inspiration of Hoefnagel's folkloric concerns, the genre scenes dominating the foreground of his townscapes suggest that the more unusual and exotic features of Iberian life interested him far more than the careful "description" of cities and towns.

For a somewhat less imaginative and definitely more chorographic portrait of urban Spain, it is necessary to turn to the work of Anton van den Wyngaerde (d. 1571), a Flemish artist who specialized in townscapes and urban views. "Of all the pleasures offered by the delightful and ingenious art of painting," he once wrote, "there is none I appreciate more than the representation of places."[26] Unlike Hoefnagel, Van den Wyngaerde was interested in the city as city, a predilection reflected in the series of drawings he prepared for Philip II, starting in 1561.

The circumstances surrounding this important commission and its relation to the Spanish monarch's other geographical projects have been examined elsewhere.[27] Yet it should be noted that the commission, in addition to whatever it says about Van den Wyngaerde's artistic merits,

FIGURE 3.4. Georg Hoefnagel. View of Seville. Originally published in Braun and Hogenberg, *Civitates Orbis Terrarum* (1572)

reflects Flemish cartography's reputation for excellence in the mid-sixteenth century as well as the many cultural and scientific contacts that existed between Spain and the Low Countries during an era when the latter still formed part of the Spanish monarchy.[28] It is not surprising then that Philip looked to one of his Flemish subjects in order to obtain what one document specifically referred to as a "painted description" of Spain's principal cities and towns, evidently in the hope of publishing some sort of a city atlas pertaining to his Iberian realms. For this purpose the king arranged to have Van den Wyngaerde's original drawings sent to Antwerp for engraving, but for reasons undoubtedly connected to the disturbances associated with the Dutch Revolt, begun in 1566, the project was never completed. In the meantime, Philip apparently used larger versions of these same drawings as wall decorations in his palaces, creating, in effect, a series of *camera delle citta* intended to demonstrate both the extent and magnificence of his Iberian realms.

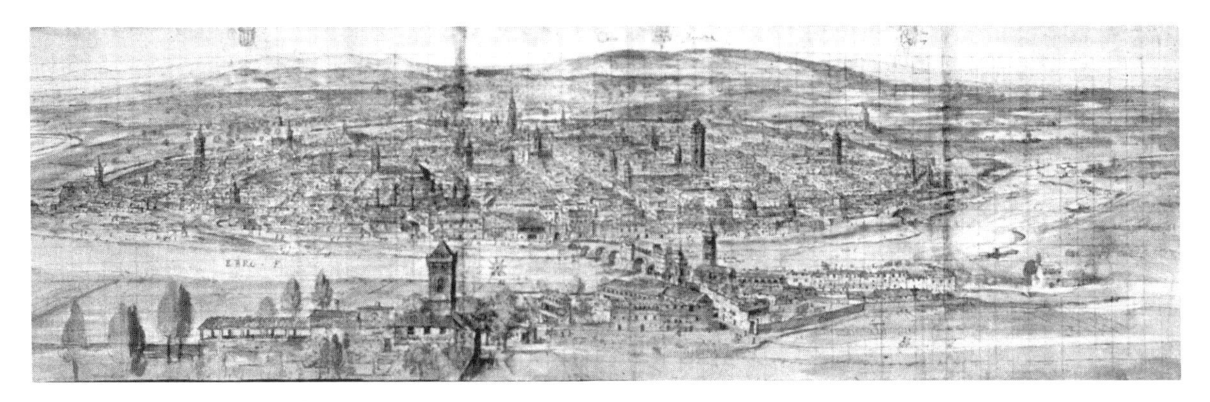

FIGURE 3.5. Anton van den Wyngaerde. View of Valencia. 1563. Nationalbibliothek, Vienna.

If judged strictly by his skills as a topographer, Van den Wyngaerde was practically without peer. In contrast to Hoefnagel, who generally demoted the city to the background of his compositions, Van den Wyngaerde placed the city closer to the forefront of his drawings and employed an oblique vantage point that allowed simultaneously for a glimpse of the street plan as well as the elevations of individual buildings. He also demonstrated his interest in the city by keeping the representation of genre scenes to an absolute minimum. In fact, many of the cities he represented appear almost depopulated, as if to reinforce his artistic interest in the city as *urbs*. Compared with the fanciful townscapes by Hoefnagel, Van den Wyngaerde's views lack life and vitality. On the other hand, they reflected his eye for architectural detail, a quality demonstrated in his preliminary site plans and sketches of individual buildings.

Yet for all their precision, Van den Wyngaerde's "painted descriptions" of Spain's cities were still not topographically accurate, that is, a "portrait" in the sixteenth-century sense of the term. In his view of Valencia, for example, he repositioned the city's cathedral for pictorial effect (figure 3.5), and generally he exaggerated the dimensions and height of churches and other religious monuments, probably to draw attention to Spain's cities as repositories of the faith, a distortion that his patron, a champion of Roman Catholicism, would have undoubtedly appreciated. Van den Wyngaerde's views also accorded emphasis to the so-called noble elements of city life: religious edifices, noble palaces, and important public buildings such as town halls, gateways, hospitals, and the like. His effort to ennoble Spain's cities also helps to explain Van den Wyngaerde's fascination with classical ruins and statues—these were generally interpreted as demonstrations of a city's grandeur and importance in times long past[29]—and also why he included in his compositions glimpses of the fields and orchards that surrounded Spain's cities together with carefully rendered images of ships, fishing boats, salt mounds, watermills, and so on. These and other motifs of urban economy not only conformed to the Aristotelian notion of the city as a self-sufficient community but also served as metaphors for the wealth and importance of the cities depicted.[30]

Van den Wyngaerde's concern with the "noble" meant that his vision of the Spanish city was idealized; crime, poverty, dirt and decay—

these all too common but definitely ignoble aspects of sixteenth-century urban life had no place in his drawings, almost as if they had been scrupulously erased. In their place he highlighted those parts of the urban fabric that contributed to grandeur and magnificence, evidently with an eye toward producing a sanitized vision of the Spanish urbs that his patron, Philip II, could proudly put on display. In this respect, Van den Wyngaerde's views are best interpreted as a simulacrum of the Spanish monarchy; they constituted emblems of the king's power, insignia of Philip's *majestad.*

Van den Wyngaerde's views are also important to the extent that his emphasis on "description" established a model followed by other itinerants and travelers, artists and cartographers alike. Most late sixteenth-century views of Spanish cities were single sheet engravings of individual cities (Seville and Toledo were the places foreign artists seemed to prefer), and the only series were the views of the towns of the Canary Islands by Leonardo Torriani (1559–1628), an Italian military engineer in Philip II's service, and the small, pen-and-ink sketches by Didacus or Diego de Cuelbis, a German traveler who visited Spain around 1600.[31]

Of these, Torriani's, consisting mostly of ground plans, are by far the most interesting. Executed between 1587 and 1592, they formed part of Torriani's *Descrittione e historia del regno de l'isole canarie,* an atlas which, in accordance with the sixteenth-century definition of "description" as an accurate or realistic portrait of a particular place, was intended to provide King Philip with a detailed account of the islands' unusual history and geography. In addition, the *Descrittione* offered Philip a blueprint for strengthening the islands' defenses—Francis Drake had raided the Canaries in October 1585.[32]

In keeping with these aims, the twenty-odd watercolor plans contained in this album—all comparable to the ground plan of Las Palmas (figure 3.6)—displayed Torriani's considerable knowledge of surveying techniques along with an economy of style that eschewed both allegory and extraneous ornamentation. The scientific and strategic goals of the *Descrittione* were also reflected in the attention paid to the city's topography, the layout of individual streets, and the disposition of principal monuments. A regular, predetermined plan even decided the color scheme: rooftops were depicted in red; squares and thoroughfares in white; the surrounding countryside in brown. Yellow was reserved for proposed defensive structures—in the case of Las Palmas, a semicircular wall, several defensive towers, and a citadel. Compared to the "ennobling" cityscapes of Van den Wyngaerde, Torriani's were positively utilitarian. In essence, they were the work of a military cartographer determined to apply, as systematically as possible, the latest techniques of his profession to the representation of the Spanish urbs.[33]

Torriani's *Descrittione* represented but one of many special-purpose city views commissioned by the Spanish Council of War. Executed for various strategic reasons, these views were tantamount to state secrets, squirreled away in the state archive and ordinarily kept hidden from public view. As a result, they were mostly workaday drawings offering a minimum of superfluous embellishment, let alone the elaborate color scheme that Torriani had employed. In this respect they resembled Holanda's sketch of San Sebastián with the exception that, from the mid-sixteenth century on, most were ground plans employing an orthogonal as opposed to oblique or equestrian perspective. As such they were primarily the work of skilled architects and military engineers, as relatively few artists possessed the mathematical and surveying skills needed to pre-

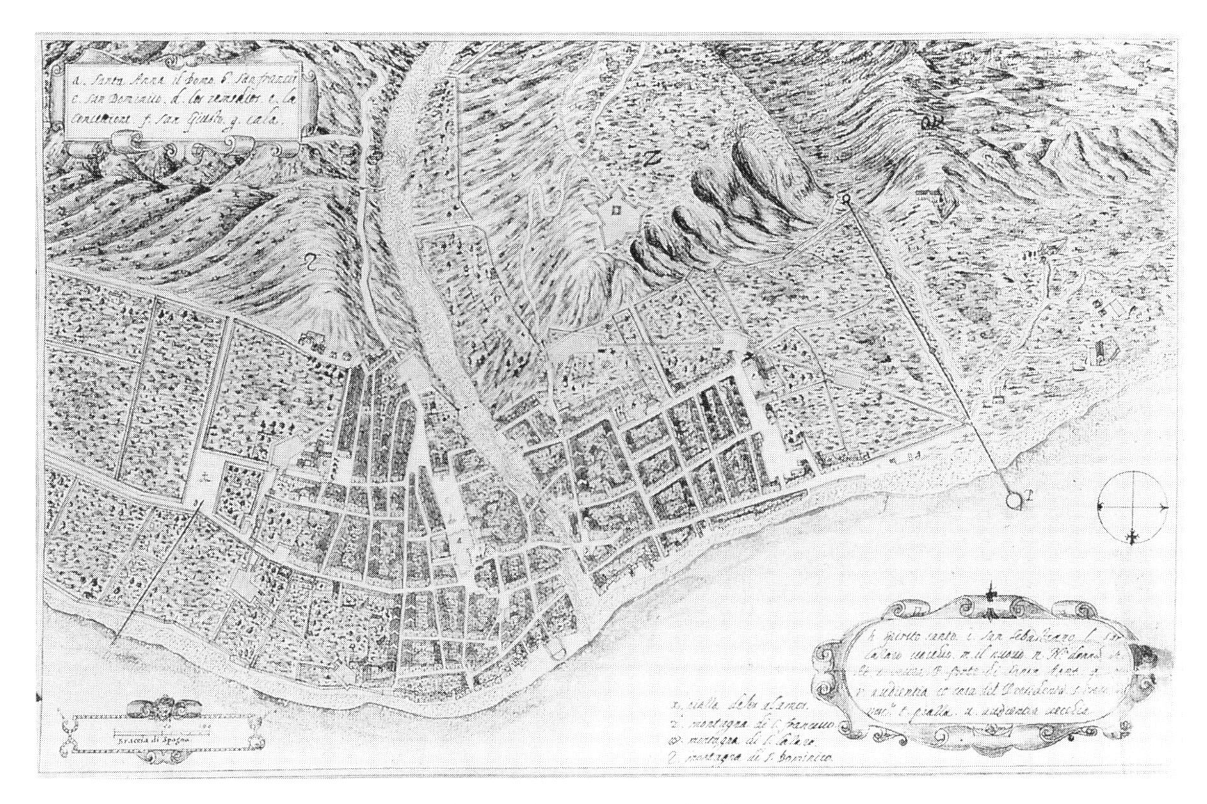

FIGURE 3.6. Leonardo Torriani. Plan of Las Palmas, Canary Islands. c. 1593

pare a detailed, accurately measured *planta* or ground plan. Such plans were pure artifice, as they depicted a city as it could not otherwise be seen. Nevertheless, they fulfilled the general precepts of chorography by offering a "true likeness" of a particular town, generally with emphasis on the city's defenses.

The first large Spanish city to have such a plan was Valencia. In 1608 that city's viceroy, the marquis of Caracena, commissioned Antonio Manceli, an Italian cartographer then resident in Valencia, to produce a ground plan that indicated not only the disposition of Valencia's streets but also highlighted the location of its principal monuments as well (figure 3.7).[34] Manceli executed his plan with the steely, measured precision of a military engineer, employing surveying techniques that were already common in sixteenth-century Italy but only beginning to find their way into Iberia. Nevertheless, this *planta* was the harbinger of things to come. Increasingly, city views that aimed at description would be done, to quote Johannes Kepler, "non tanquam pictor, sed mathematicus" (not in the manner of a painter but in that of a mathematician or scientist).[35]

On completing his work in Valencia, Manceli headed for Madrid where, in 1622, he completed a ground plan of the *villa y corte*, the town's first. This plan, known commonly as the De Witt plan, anticipated the monumental *topographia* of Madrid completed by the Portuguese cartographer

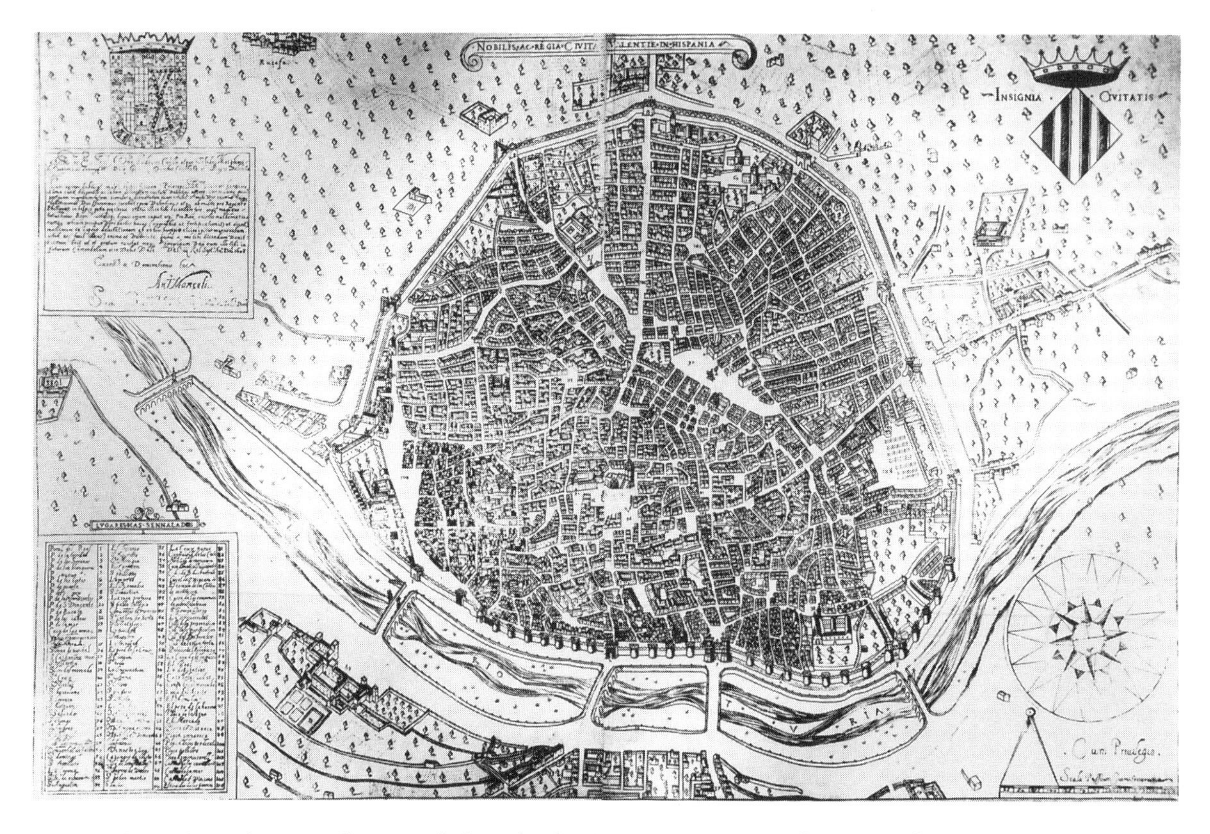

FIGURE 3.7. Antonio Manceli. Ground Plan of Valencia. 1608. Private Collection, Valencia, Spain.

Pedro Teixeira and published in Antwerp in 1656 (figure 3.8).[36] Technically, the *topographia* did not offer a true orthogonal plan of Madrid as Teixeira included within it building elevations and other details that could not be seen from a ninety-degree perspective. Nor was it wholly topographically accurate. Teixeira claimed that he executed the plan "True to life, so that one can count the doors and windows of each [building]," yet he exaggerated the dimensions of certain buildings, especially those connected with the Buen Retiro, Philip IV's new palace, in order to glorify the monarch to whom the map was dedicated. Royal symbolism was also evident in the inscription, "Mantua Carpe[n]tanorvm sive Matritvm Urbs Regia" (Mantua of the Carpentana, or Madrid, Royal City), the Habsburg shield, and the pedestal surrounded by the spoils of war evidently intended to emphasize Philip's military power and strength. Similarly, the pedestal's inscription—Philippo IV / Regi Catholico / Forti et Pio / Urbem hanc suam / et in ea orbis sibi subiecti / compendium / exhibet MDCIII [*sic*] (This [map] displays to Philip IV, Catholic Monarch, Strong and Pious, this, his city, and summation of the world subject to him)—emphasized Madrid as the capital of Philip's global monarchy. For all of its apparent precision, therefore—or, as Teixeira expressed it, its being "drawn *al natural*"—the *topographia* offered a symbolically charged view

FIGURE 3.8. Pedro Teixeira. *Topographia* of Madrid. 1656

that highlighted Madrid's connection to the crown. In this respect the *topographia* emulated the "monarchocentric" map view of Paris done by Bénédit de Vassalieu (dit Nicolay) for Henry IV in 1609.[37]

Despite the increasing tendency in the seventeenth century for chorographic views to utilize advanced mapping and surveying techniques, the "description" of cities did not necessarily require, as Kepler had expressed it, a *mathematicus*.[38] City views, as Vermeer's *View of Delft* readily suggests, remained within the artist's domain. In Spain, one such descriptive view is the pen-and-ink drawing of the Cantabrian port of Santander attributed to the Dutch artist Bonaventura Peeters. In this instance the artist, apparently working without the help of instruments, offered an impressionistic yet topographically accurate view of the city as seen from a vantage point at sea (figure 3.9).[39] Much the same is true of the cityscapes of Pier Massimi Baldi, the Italian artist who executed profile views of more than thirty Spanish cities as he accompanied Cosimo de' Medici on his visit to Spain and Portugal in 1668.[40] These views suggest that a "true likeness" of a city did not necessarily always take the form of a map.

This particular emphasis is important for understanding the *View of Zaragoza* (1647) by the court artist, Juan Bautista Martínez de Mazo

FIGURE 3.9. Bonaventura Peeters. View of Santander. Seventeenth century. Scheepsvaart Museum, Amsterdam.

(figure 3.10).[41] Mazo, Diego de Velázquez's son-in-law, accompanied King Philip IV and his son, Prince Baltasar Carlos, on a journey to Zaragoza, capital of Aragon, in 1646. The prince became ill and soon died, and it was possibly to commemorate this sad occasion that the monarch commissioned Mazo to paint a view of the city. The painting subsequently entered the royal collection and won praise for its "precision" (*puntualidad*) from Antonio Palomino, a late seventeenth-century critic and historian of Spanish art.[42] Palomino's term applies perfectly to Mazo's sweeping view of Zaragoza, at the center of which is a startlingly frank and realistic depiction of the ruins of the city's principal bridge across the Ebro, which had been destroyed during a flood in 1643. Absent is the sense of grandeur and nobility that Van den Wyngaerde had imparted to this city in his view of Zaragoza, eighty years earlier.[43] The image is rather that of a city in decay, almost as if Mazo purposely set out to produce a portrait of Zaragoza which, however somber and stately its mood, still qualified as "description."

COMMUNICENTRIC VIEWS

IF THE DESCRIPTION of *urbs* represented the primary aim of chorographic views, "communicentric" views were more interested in representing *civitas,* the other facet of Isidore's definition of city. It follows that these views, many of which never moved much beyond the community that they were originally intended to represent, offered a different image of the city than those prepared for "descriptive" purposes.[44]

Although there are several medieval views of Spanish cities that fall within the category of communicentric view, an excellent sixteenth-

FIGURE 3.10. Juan Bautista Martínez de Mazo. *View of Zaragoza.* 1647. Prado Museum.

century example is the representation of Orihuela included as a full-page illustration in a manuscript, dated 1578, listing the city's fiscal and juridical privileges (figure 3.11).[45] In the illustration, a city of exaggerated and imaginary proportions looms up in the distance where it serves as backdrop for a battle—evidently, Orihuela's resistance to a lengthy siege ordered by Pedro I of Castile in 1364—raging below. Although the view contains some identifiable monuments, notably the medieval castle and the collegiate church of San Salvador, it is primarily a *typus* or conventionalized portrait incorporating stock elements of an urban landscape: walls, gates, densely packed houses, and so on. As noted above, such conventional images were relatively commonplace in the early sixteenth century, but in this instance they suggest that the artist was less interested in "describing" Orihuela than in illustrating the lengths to which its citizens went to defend their communal privileges under the watchful eye of their heavenly advocates, Saints Justa and Rufina. In other words, what seems to define Orihuela as Orihuela in this image is not the city's physical properties but rather the memory of the battle in which *oriolanos* defended the laws and privileges that made their city unique.

Messages of a similar kind pervade most other communicentic views, among them the *retrato* or portrait of Burriana—one of several similar views—included in Martín de Viciana's history of the kingdom of Valencia, first published in 1563 (figure 3.12). The style is awkward, even naive; nevertheless this "portrait," evidently the work of Viciana, a native of Burriana, attempts to particularize this conventionalized view by highlighting those monuments that were particularly

FIGURE 3.11. Anonymous. View of Orihuela. 1578? Archivo Histórico Nacional, Madrid, Codices, 1368b.

FIGURE 3.12. Martin de Viciana? View of Burriana. 1563.

valued by local inhabitants. These included the town's encircling walls and moat; the gate through which Jaime IV, the monarch responsible for Burriana's reconquest from the Moors, first entered the town; the parish church, described in the accompanying text as "a very large and beautiful temple;" and, *extramuros*, the hermitage of San Mateo, "a house of much devotion and well-frequented by devont Christians."[46] In other words, Viciana's portrait offered less of a "true likeness" of Burriana than an image meant to provide the book's readers (most of whom were Valencians) with a view focusing on those monuments that commemorated the community's history and its faith.

The view included in Luis de Toro's "Description of the City and Bishopric of Plasencia" (1573) served similar ends, albeit with a measure of topographical precision approximating that of some chorographic views. Toro, a physician, prepared this manuscript volume to provide Plasencia's bishop-elect with advance information about his new diocese, and did so in the inflated language of a tourist guide. Accompanying his "description" was a map portraying the diocese as the heavens in miniature: a series of concentric circles in which the city of Plasencia occupied the symbolic center. The accompanying view of the city of Plasencia was similarly tailored for maximum pictorial and symbolic effect (figure 3.13). Adopting a panoramic format similar to that utilized by Hoefnagel and Van den Wyngaerde, Toro—who was presumably the author of the view—not only made the city appear much more extensive than it actually was but also wildly exaggerated the size of the city's cathedral, ostensibly to impress the new prelate

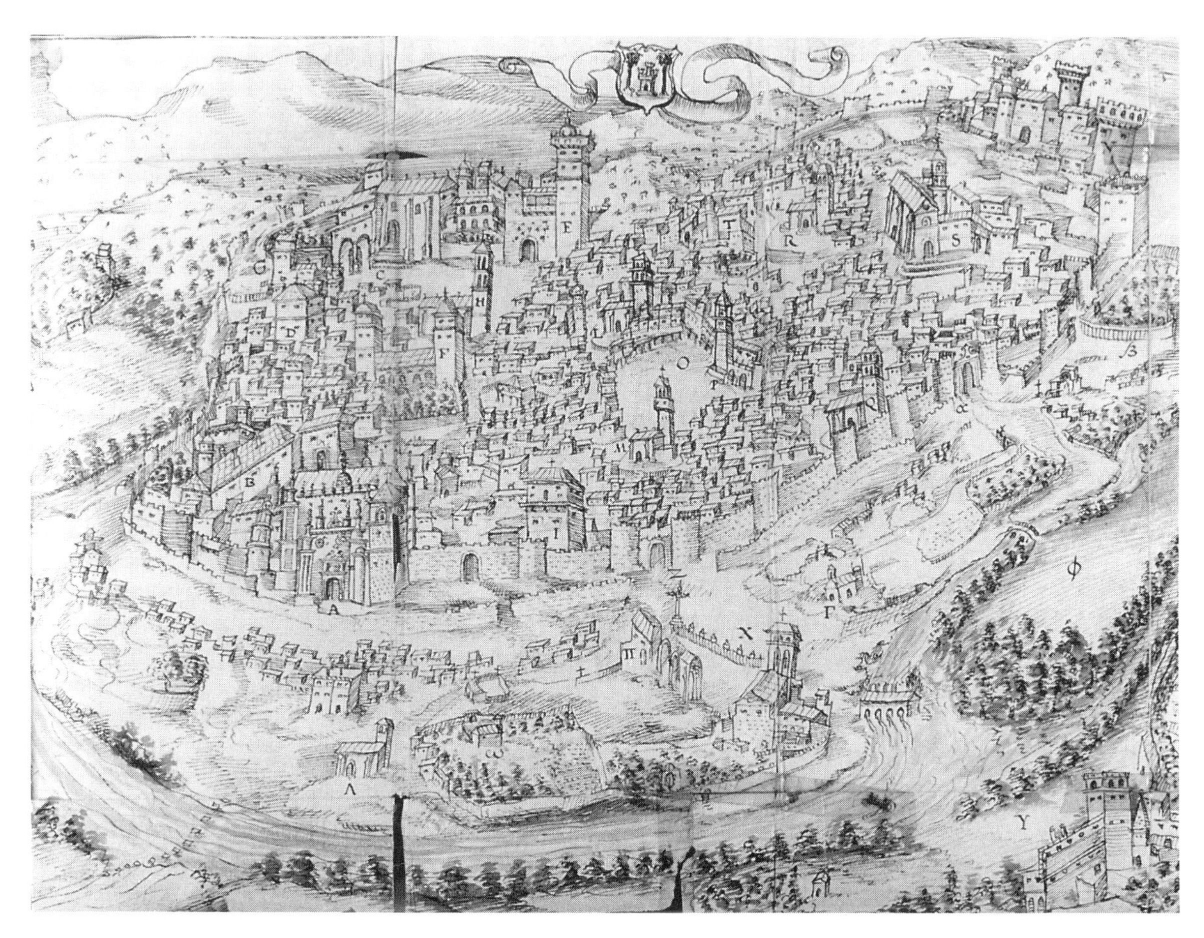

FIGURE 3.13. Luis de Toro? View of Toro. 1573. Ms., University of Salamanca.

with its grandeur and magnificence and in so doing to honor the community to which he belonged.[47]

Self-promotion of this sort is characteristic of most other communicentric views, although relatively few of these images, in the sixteenth century at least, adopted the panoramic format utilized by Toro. In this respect Spain differed from Italy, the Holy Roman Empire, and the Low Countries, where the demand for panoramic cityscapes seems to have been increasing steadily.[48] In contrast, Spanish city dwellers generally made do with images that represented their cities in other ways, generally by means of a shorthand technique in which a single structure—a signature building—served as the emblem or icon of the city as a whole.[49] This particular method is evident in the frontispiece in an incunabular edition of Eiximenis's *Regiment de la cosa publica* published in Valencia before 1500 (figure 3.14). In this woodcut, the Puerta de Serranos, Valencia's principal gateway and the site where criminals were customarily punished, symbolized Valencia's *res publica,* the physical embodiment of its

FIGURE 3.14. Anonymous. Puerta de Serranos. Woodcut originally published in Francesc Eiximenis, *Regiment de la cosa publica* (Valencia, c. 1490)

ayuntamiento or civic government.[50] Valencia's other signature building, the Miquelet, the late fourteenth-century campanile which dominated the city's skyline, served a similar purpose, especially for artists inclined to represent the city in spiritual terms.

Other structures used to identify individual Spanish cities included Segovia's Roman aqueduct, or *puente,* and Seville's Giralda, the late twelfth-century minaret that was subsequently transformed into the cathedral's bell tower. A Giralda of exaggerated proportions figured prominently in most sixteenth-century chorographic views of Seville, notably in those by Hoefnagel, who additionally provided readers of the *Civitates* a cutaway image of this important structure.[51] An outsized and divinely illuminated Giralda also dominated the anonymous late sixteenth-century view of Seville as seen from Triana (currently housed in Madrid's Museo de América).

Meanwhile, Sevillian view makers, customarily eschewing the panoramic format preferred by visiting artists, fixed on the Giralda as an icon of the city, generally depicting it in isolation as if no other structure was necessary to conjure up a mental image of Seville as a whole.[52] The tower's frequent representation in various media, including sculpture, ceramics, and painting, further attested to the tower's iconic function (figure 3.15). Thus, in the same way that the Parthenon came to symbolize Athens, the Coloseum Rome, and the Eiffel Tower Paris, the Giralda became Seville.

More specifically, the Giralda carried with it certain religious connotations that in effect accorded Seville a special relationship with God. The tower's spiritual significance dated from at least 1396 when it was reported that Saints Justa and Rufina had appeared miraculously during a violent windstorm in order to protect the Giralda from harm.[53] The figure of Faith, known popularly as the Giraldillo, set atop the tower in 1568 further enhanced the Giralda's architectural prominence within Seville, but it also embodied the idea of Christianity's triumph over Islam. The Giralda thus represented Seville's particular contribution to the Faith Militant and linked the city to one of the central elements of the Counter Reformation Church.

Metaphorically, therefore, the Giralda transformed Seville into *civitas dei,* a heavenly city dedicated to the service of God. This in fact was

FIGURE 3.15. Anonymous. Ceramic tile with the Giralda with Saints Rufina and Justina. Sixteenth century.

precisely the image of the city that the prominent Sevillian painter Francisco Pacheco (1564–1654) attempted to convey when he used Seville as a backdrop for various representations of the Virgin of the Immaculate Conception (figure 3.16 and 3.16a). The attributes of the *Inmaculada* are both many and complex, but they generally include the Tower of David and the Heavenly City.[54] In Pacheco's painting the Giralda became this sacred tower, and Seville the equivalent of the City of God, a theme that repeated itself in a later rendition of this same subject by Francisco de Zurbarán (1598–1664) that was apparently intended to be hung in Seville's town hall.[55] Although Seville is clearly recognizable in these paintings, the emphasis was less on the city as *urbs* than as *civitas,* in this case a *civitas cristiana,* a Christian community wholly devoted to the doctrine of Mary's immaculate birth, another Counter Reformation ideal.

Christian iconography of a similar kind infused most other seventeenth-century communicentric views of Spanish cities. This designation certainly applies to El Greco's early seventeenth-century *View of Toledo* (Metropolitan Museum of Art), a painting that blatantly sacrificed topographical accuracy in order to highlight the city's role as Spain's spiritual capital (figure 3.17). In this case, the artist, a longtime resident of Toledo, deliberately moved the cathedral from its actual location to the center of his composition, where it stands atop a high promontory overlooking the valley of the Tagus River. El Greco also distorted Toledo's urban landscape by including at the left of the painting what appears to be a monastery sitting on top of a cloud. This is apparently a reference to Agaliense monastery, a building that had disappeared by the seventeenth century but one whose precise location local clerics debated because of its association with Saint Ildefonso, the seventh-century Toledan bishop known for his treatise propounding the doctrine of the Virgin's immaculate conception.[56]

El Greco, however, was not the only artist willing to alter topographical realities in order to convey certain religious or spiritual messages. Much the same can be said about Diego de Astor, a well-known engraver whose image of Santiago de Compostela originally appeared in Mauro Castello's *Historia del Apostol Santiago,* published in 1614. This book was a lengthy defense of Saint James's arrival, preaching, and subsequent burial

FIGURE 3.16. Francisco Pacheco. *Virgin of the Immaculate Conception.* Seville.

FIGURE 3.16a. Detail from figure 3.16 (showing Torre de Oro, Giralda, etc.)

in the city then called Iria Flabia,[57] and Astor's engraving represented it as a perfect square: a New Jerusalem built according to Roman principles of urban design and set in the middle of a landscape studded with holy sites associated with the apostle.

Ambrosio de Vico's remarkable *plantaforma* of Granada, executed in 1612, represents yet another and even more ambitious attempt to create an idealized Christian city. Vico was an architect of Italian origin who settled in Granada in 1575 and subsequently intervened in various architectural projects initiated in the 1580s by Pedro de Castro, the city's archbishop.[58] It was Castro in fact who commissioned the *plantaforma* as part of a carefully orchestrated effort to enhance Granada's image as a city of Christian heritage and design. Toward this end Castro and subsequent Granada archbishops organized an elaborate propaganda campaign designed to prove the authenticity of the *plomos de Sacramonte,* a series of lead boxes containing parchments miraculously "discovered" in 1595 and which purportedly shed light on Granada's conversion to Christianity in the first century A.D. Castro also commissioned several new ecclesiastical histories of the city that studiously ignored eight centuries of Muslim rule in order to glorify Granada's Christian heritage.[59]

The *plantaforma* constituted yet another facet of this larger project. But what appears to be a

FIGURE 3.17. El Greco. *View of Toledo.* C. 1610. Metropolitan Museum of Art.

topographically accurate ground plan of the city is actually little more than an idealized projection of what a fully Christianized Granada might have been like (figure 3.18). Urban reforms executed since 1492 by Granada's archbishops, Castro among them, had resulted in the new cathedral and several open plazas of Renaissance design, but much of the city's Muslim fabric—a maze of narrow, twisting streets and cul de sacs—remained untouched, particularly in the northeastern section of the city known as Albaicín.[60] Yet in Vico's ground plan the streets of this particular quarter are straightened and cul de sacs replaced by squares in an effort to portray Granada as a city of modern, that is to say, Christian design.[61] In other words, what the *plantaforma* depicted was the city of Castro's dreams. Vico transformed Granada's Muslin *urbs* into a *civitas cristiana* just as El Greco had converted Toledo into a spiritual powerhouse, a city radiating a special kind of religious power.

In the seventeenth century, the concept of the city as a spiritual community was practically synonymous with the image that most of Spain's cities sought to project. It pervaded, for example, the many city histories published during this era. As elsewhere in Europe, these histories were often little more than antiquarian panegyrics, intended principally to enhance a city's nobility by calling attention to its remote origins, its importance in classical times, its early charters and other privileges. Centuries of Muslim domination, however, often made this quest for nobility elusive. Consequently, many local historians, attempting to erase from the historical record all traces of Muslim influence, emphasized their city's long Christian heritage by portraying it as a crucible of Catholicism, mother of martyrs and saints, and staunch defender of the faith. This particular effort to recover a city's Christian heritage can be partly explained by the fact that the authors of these histories tended to be clerics with a vested interest in religious history, but even secular authors were apt to conceive of their city as a *ciudad de Dios.*[62] The idea of cloaking the city in spiritual garb can also be connected to the desire of the Habsburg monarchy to portray itself as the champion of Roman Catholicism as well as to the widely held belief that Spaniards were divinely ordained to nurture and to protect the True Church. Municipal chroniclers were eager to prove that their community played an important part in this divine mission and consequently constructed a history compatible with the notion of both early and eternal adherence to the Catholic faith. Meanwhile, local historians borrowed heavily—and often uncritically—from the notorious "false" chronicles of Flavius Dextro and Luitprando, two alleged eyewitnesses to

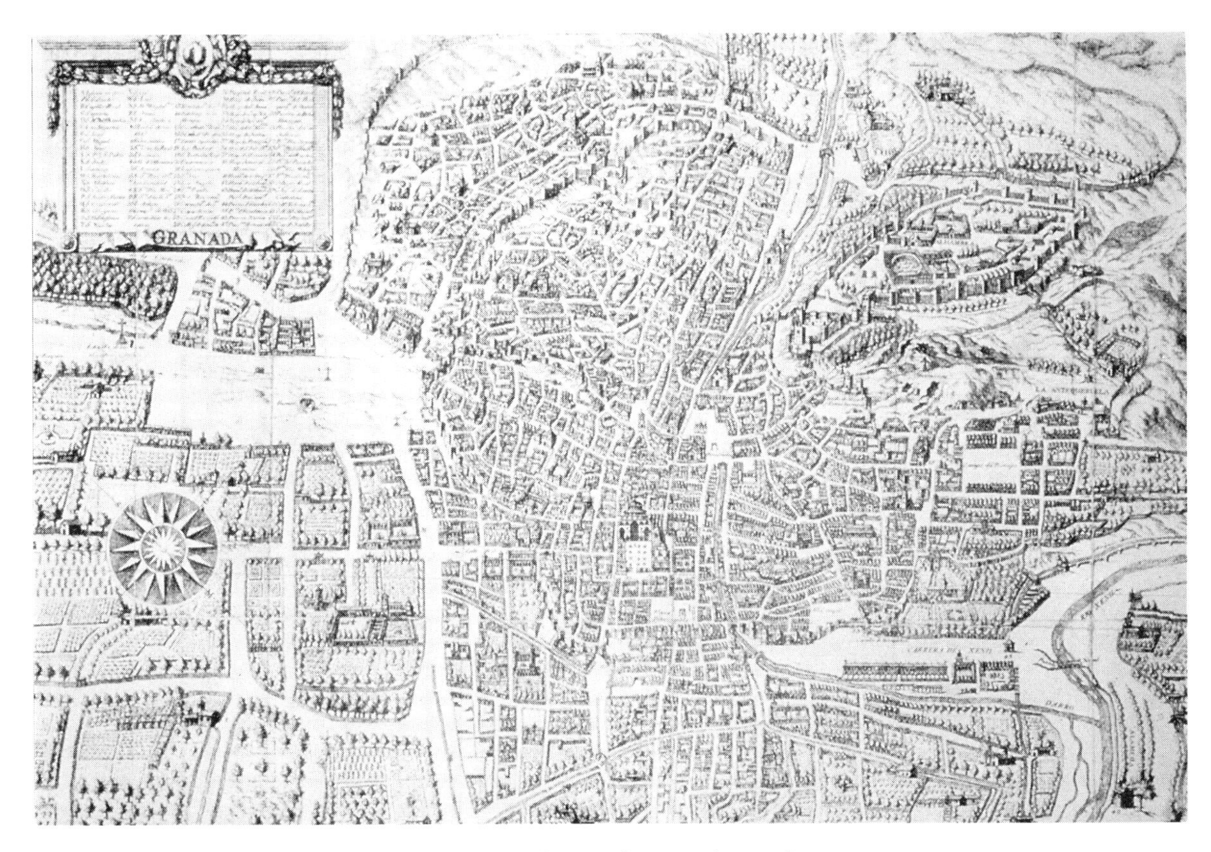

FIGURE 3.18. Ambrosio de Vico. Plantaforma of Granada. Late sixteenth century.

Spain's early Christian history, in order to document such momentous "historical" events as the arrival of James the Apostle in their city as well as to recover the names of their first bishops, their earliest martyrs, and other holy figures.[63] This desire to construct and, if necessary, to fabricate a fully Christian history also led scholars to elaborate their city's contributions to the success of the reconquest, the monarchy's efforts to defeat Protestantism, and the church's struggle to impose Christianity on the inhabitants of the New World. So pervasive, in fact, was this sacred or spiritual conception of urban history that other ways of structuring and thus ennobling a city's past, by examining its architectural heritage, for example, or by tracing the history of its laws and governing institutions, attracted only minimal attention on the part of local historians.[64]

A spiritual orientation of a related kind infused most seventeenth-century views of Spanish cities, the majority of which were executed for devotional purposes and then in conjunction with the worship of particular saints of local importance. Cuenca, a city in La Mancha, is a case in point. With one exception—a painting, now lost, documenting the celebrations that city staged in honor of Philip IV in 1642—the only

FIGURE 3.19. Eugenio Cajes. *Saint Julian of Cuenca.* C. 1630. Pollok House, Glasgow.

FIGURE 3.20. Anonymous. *Virgin of the Mercé with View of Barcelona* C. 1650. (Destroyed)

seventeenth-century views of Cuenca are those included in the background of paintings devoted to the life of Saint Julián, the city's patron saint.[65] Of particular interest in this regard is Eugenio Cajes's *Saint Julian of Cuenca* (Pollok House, Glasgow) (figure 3.19), a portrait that includes a glimpse of the impressive stone bridge that spanned the Huécar River and which then served as Cuenca's signature building.[66]

City views appended to various paintings devoted to the Virgin Mary served similar devotional purposes. In Barcelona, for example, as in Seville, chorographic views of the city were primarily the work of foreign artists; views by local artists tended to emphasize the city as a spiritual community or *civitas dei.* One such view appeared in a painting, formerly in the Church of the Mercé (and destroyed during the Spanish Civil War), in which Barcelona's municipal counselors, the elected representatives of its *civitas,* kneel before a Madonna, evidently to thank her for having helped the city during a recent plague. Beneath these figures the artist inserted a view of Barcelona as seen from the heights of Montjuic (figure 3.20). In this votive painting, however, the view of Barcelona's *urbs,* for all of its appar-

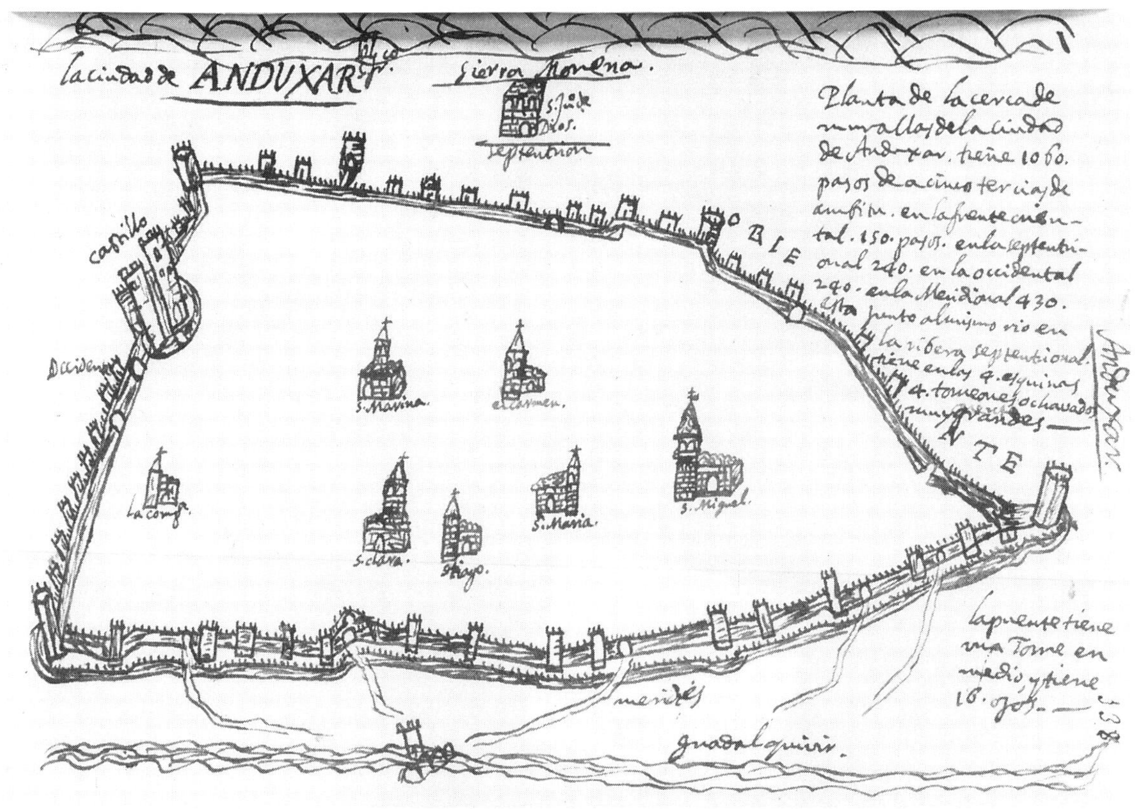

FIGURE 3.21. Anonymous. Plan of Andújar. C. 1630. Biblioteca Nacional, Madrid.

ent concern with "description," served principally as a metaphor for its *civitas* and to underscore the citizenry's devotion to Mary herself.[67]

The theme of the city as a spiritual community united by faith is in fact common to most seventeenth-century Spanish city views. It appears, for example, in a drawing of the Andalusian town of Andújar included in a manuscript history of the diocese of Jaen dating from around 1640 (figure 3.21). As the map of the diocese included in the history suggests, the clergyman responsible for the view evidently had some cartographical training and was therefore capable of representing Andújar from an elevated, oblique perspective. Yet the view was highly selective. It emphasized the town's churches and encircling walls but made no reference to any secular buildings let alone the urban fabric itself. Voids such as this often appeared in ground plans intended for military purposes, which understandably emphasized fortresses, walls, and other defenses but often left the rest of the city a complete blank. In this instance, however, the void seems to have been connected to the chronicler's effort to represent Andújar as a holy community wholly committed to the defense of the church.[68]

A similar message pervades the images commissioned at the time of the fiestas held in Valladolid in 1656 in honor of that city's leading penitential confraternity, the Cofradía de la Vera

FIGURE 3.22. Felipe Gil de Mena? Silverith's Street and Church of the Holy Cross, Valladolid. 1656. Private Collection, Madrid.

Cruz. These celebrations, spread over the course of four days and centered on a series of processions staged in and around Valladolid's main square, were a roaring success, evidently uniting much of the local populace in what was tantamount to a collective demonstration of faith, an auto-da-fé albeit one without heretics awaiting punishment. An artist, possibly Felipe Gil de Mena (1603–73), recorded the festivities in a series of four paintings, each devoted to a different day of the celebrations and depicting different parts of Valladolid's physical fabric—the street of the Platería and the church of the Holy Cross appears in figure 3.22. Together, the paintings constitute an invaluable source for reconstructing the architecture of seventeenth-century Valladolid, but description of the city's *urbs* does not appear to have been the artist's main intent. Rather, he seemed more interested in capturing Valladolid at the moment during which its citizens, through their collective act of piety, had miraculously transformed themselves into a *civitas cristiana,* a city wholly united by faith.[69]

FIGURE 3.23. Anonymous. *Triumph of San Fernando.* 1672.

The beatification of Fernando III of Castile, the thirteenth-century monarch responsible for the conquest of much of Andalusia from the Moors, led to a similar series of images for Seville. In this case, news of Fernando's beatification arrived in Seville in 1671 and served as the catalyst for a series of commemorative engravings and paintings of the saint.[70] One of these, *El Triunfo de San Fernando* (figure 3.23) depicted the monarch receiving the keys of the city from its defeated Muslim rulers. The background offered a view of Seville, but the principal aim of the painting was to offer a symbolic representation of the reconquest and the triumph of Catholicism over Islam.

Other communicentric views were perhaps less single-mindedly religious than this. One interesting example is the view of Béjar, a late seventeenth-century painting commissioned by the town's lord, the duke of Béjar, and which is still in that family's possession. In this view *urbs* is depicted, with special prominence accorded to the ducal palace, the front of which, in the manner of Velázquez's *Venus and Cupid* (National Gallery of Art, London), may be seen reflected in a mirror (figure 3.24 and 3.24a). Yet this particular painting was not expressly intended to celebrate Béjar's physical grandeur. Rather, the artist, highlighting the bullfights and other festivities that marked the community's annual fiesta, seemed intent on emphasizing the beneficence of the duke's seigneurial rule. Other nobles may have commissioned paintings of a similar sort, but on the whole the preferred mode of representing towns and cities in seventeenth-century Spain was to portray the community as a repository of faith.

CONCLUSION

SUMMARIZED HERE then is a Spanish tradition of city views in which description, with its emphasis on topographical specificity—the city as *urbs*—was considerably less developed than that of representing the particular qualities that rendered each city and its inhabitants—the city as *civitas*—unique. This kind of moralized geography, as Jürgen Schulz has argued, was typical of most medieval city views and many of Renaissance origin as well.[71] By the mid-sixteenth century, however, this particular tradition was rapidly being replaced by chorographic views offering a more or less accurate "description" of a particular place.

The "art of describing," to borrow Svetlana

FIGURE 3.24. Anonymous. *View of Béjar.* Late seventeenth century. Private Collection, Seville.

Alpers's trenchant phrase, first appeared in Spain in the guise of Anton van den Wyngaerde and was subsequently perpetuated by a series of foreign-born military engineers such as Torriani, Manceli, and Teixeira. The first Spaniard to adopt this particular style was possibly Antoni Garau, the mathematically trained cleric who executed a detailed, axonometric view of Palma de Mallorca in 1644, but Garau remained something of an exception (figure 3.25).[72] In most parts of the kingdom, description did not make much of an appearance before the start of the eighteenth century, and, even then, it required another half century or so until it became an important part of the Spanish tradition of city views.

What explains this delay? Did Spain lack the mathematical and surveying skills necessary to work in the new style? One scholar has recently argued that Spain's failure to develop a native school of terrestrial cartography meant that Spanish mapmaking after the mid-sixteenth century was almost totally dependent on the skills of other nations: at first the Portuguese, then the Italians, and, increasingly, the Flemish.[73] But there was never any lack of either native-born architects or engineers, most of whom, presumably, had the training required to produce either a detailed ground plan of a city or an urban panorama drawn to scale.[74]

In other words, the relative paucity of views

FIGURE 3.24a. Detail from figure 3.24 (with palace of the dukes of Béjar).

offering a "true likeness" of the Spanish city as *urbs* cannot—and indeed, should not—be attributed simply to a shortage of cartographic talent. It seems rather to have been a question of demand, possibly the result of indifference or lack of concern among civic and religious leaders, that is, those individuals who in other European cities were generally responsible for commissioning descriptive views and other images that celebrated the city as *urbs.* In seventeenth-century Spain, however, these same individuals—municipal councillors, local noblemen, bishops, and other prominent clergymen—seemingly disdained "description."[75] What they wanted instead was an idealized image—the city as imagined, rather than the city-as-seen, or, at the very least, a public image of their city that was consistent with the Christianized image of urban grandeur and perfection presented in the local histories that they themselves had helped to sponsor and produce.

In the end, there is something ironic about the Spanish tradition of city views. Seventeenth-century Spain is a culture known for the picaresque, a literary genre that can be interpreted as one of the most frank and realistic portraits of urban life ever written. Yet this same culture apparently expressed little interest in producing visual "descriptions" of the same cities in which Guzmán de Alfarache, El Buscón, and other well-known *pícaros* played most of their tricks. Having read such novels, where the city appeared as a New Babylon, brimming with crime and depravity of every sort, were Spain's civic and religious leaders reluctant to see themselves as they really were? Were they unwilling, so to speak, to look at themselves in the mirror, to confront economic and social realities which they might otherwise seek to avoid? The seventeenth century was not especially propitious or prosperous for most of these cities. The majority, in Castile as well as in Aragon, were experiencing a protracted period of demographic decline and economic decay from which there was little relief.

Under these circumstances, it is perhaps not surprising that local artists and draftsmen preferred to ignore present realities in order to depict their cities more imaginatively and in ways that evoked some larger spiritual reality or harked back to an earlier, more glamorous moment in their communities' history. As a result, the connection between locally produced images of cities and chorography remained tenuous. Chorographic views were generally "descriptions" seeking to convey a "true likeness" of a particular place. They focused on the city as *urbs.* Communicentric views did not necessarily blot out or distort a city's physical reality, but the urban image they proffered tended to be both

FIGURE 3.25. Antoni Garau. *Plan of Mallorca.* 1644. Private Collection.

highly selective and metaphorically charged. In other words, *urbs* in these views served mainly as a screen; peering through it afforded a glimpse of the city's inner self, the city as *civitas.*

In conclusion, I want to make clear that these distinctions are not hard and fast. In other countries, and in other eras, the criteria that appear to separate the chorographic from the communicentric city view in sixteenth- and seventeenth-century Spain may not exist. In the eighteenth century, for example, the widespread dissemination of the surveyor's art meant that communicentric views looked deceptively like chorographic views, especially in terms of the cartographic techniques they employed. Functionally, however, the two ways of depicting cities remained quite distinct, both in terms of audience and design. Chorographic views, so many of which were incorporated in published atlases and other geographical compendia, were, like the illustrations in today's tourist brochures, intended primarily for a public living outside the place they depicted. In contrast, communicentric views served a home audience that was presumably already endowed with a "public image" of their city or town. Their purpose was to foster and to strengthen ties of community as well as to offer that community ways of understanding the collective devotions and traditions that set their city apart from others. Such goals were always elusive, but in the cities of seventeenth-century Spain they saddled local artists and view makers, not to mention the local historians, with two seemingly impossible and often contradictory tasks. One was the conversion of decay into dignity. The other was to transform what visitors saw as cities of stone into what local inhabitants viewed as cities of God.

Notes

Various drafts of this essay were read by James Amelang, Henry A. Millon, Orest Ranum, Marianna Shreve Simpson, and the members of "The Seminar" in the Department of History of the Johns Hopkins University. I am grateful for their comments and suggestions, both stylistic and substantive.

1. As translated in Chiara Frugoni, *A Distant City: Images of Urban Experience in the Medieval World,* trans. William McCuaig (Princeton, N.J., 1991), 3. Isidore's definition of city follows that of Aristotle, *Politics,* 1.5–14, 3.9 as well as Saint Augustine, *City of God,* 15.8.

2. Giovanni Botero, *The Reason of State and the Greatness of Cities,* ed. P. J. Whaley and D. P. Whaley (London, 1956).

3. *Partida* 7, *título* 33, *ley* 6 describes a town as "todo aquel lugar q es cercado de los muros, con los arrabales et los edificios q se tienen con ellos," whereas *Partida* 2, *título* 10, *ley* 1 reads: "Pueblo llama el ayuntamiento de todos los omes comunalmente, de los mayores e los medianos, e de los menores."

4. I refer here to Francesc Eiximenis, *Tractat de regiment de princips et comunitas* (Barcelona, 1904), the twelfth book of his great encyclopedia on Christian morality, *Lo Chrestiá,* and to Fray Juan García de Castrojérez, *Glosa castellana,* ed. Juan Beneyto Pérez (Madrid, 1947), which is actually a translation and gloss of Egidius Romanus, *De Regimine Principe.* For later writers, see Antonio Antelo Iglesias, "La ciudad ideal según Fray F. Eiximenis y Rodrigo Sánchez de Arevalo," *La ciudad hispánica* (Madrid, 1985), 1:19–50; Fray Alfonso de Castrillo, *Tratado de la Republica* (Burgos, 1521), 3v; Diego Pérez de Mesa, *Política y razón de estado* (1623), ed. L. Pereña y C. Baciero (Madrid, 1980), 11–19. For a broader discussion of the idea of the city in early modern Spain, see Santiago Quesada, *La idea de ciudad en la cultura hispana de la edad moderna* (Barcelona, 1992).

5. Dámaso de Frias, "Diálogo en alabanza de Valladolid," in N. Alonso Cortés, *Miscellánea vallisoletana* (Valladolid, 1955), 2:251. The original reads: "una congregación de muchas familias conformes en leyes y costumbres con fin de abundar en todas las cosas necesaria de vida."

6. Sebastián de Covarrubias, *Tesoro de la lengua castellana o española* (Barcelona, 1943), 427.

7. "Description" meant to write about or to represent something without exaggeration and with an exacting degree of verisimilitude. Covarrubias, ibid., 457, states: "Descrevir. Narrar y señalar con la pluma algun lugar o caso acontecido, tan al vivo como si lo dibuxara. Descripción, la tal narración o escrita o delineada, como la descripción de una provincia o mapa." For more on the term's meaning, see Svetlana Alpers, *The Art of Describing: Dutch Art in the Seventeenth Century* (Chicago, 1983), 136.

8. Claudius Ptolemy, *Geography,* trans. Edward L. Stevenson (New York, 1932), bk. 1, chap. 1.

9. Petrus Apanius, *Libro de cosmographia* (Antwerp, 1548), chap. 4.

10. My understanding of "communicentric" views derives in part from Barbara Mundy, *The Mapping of New Spain* (Chicago, 1996), 116.

11. For a recent discussion of the program of Lorenzetti's fresco, see Loren Partridge and Randolf Starn, *Arts of Power: Three Halls of State in Italy, 1300–1600* (Berkeley, 1992). The city views painted in the town halls of various German cities in the sixteenth century served similar purposes. See Kristin Eldyss Sorensen Zapalac, *"In His Image and Likeness": Political Iconography and Religious Change in Regensburg, 1500–1600* (Ithaca, N.Y., 1990).

12. Kevin Lynch, *The Image of the City* (Cambridge, 1960), 7.

13. The classic treatment of this genre's development is James Elliott, *The City in Maps: Urban Mapping to 1900* (London, 1990). For the early modern era, see Jürgen Schulz, *La cartografia tra il arte e la scienze* (Ferrera, 1990), and Lucia Nutti, *Ritratti di città: Visione e memoria tra medioevo e settecento* (Venice, 1996).

14. Archivo General de Simancas: Mapas, Planos y Dibujos X-1. The street in question was the calle Barrionueva,

located in the northeast of the plan. To allow it to reach the town's central plaza in a straight line called for the demolition of several houses located on the calle de Pozo. The owners of these houses objected, challenging the right of the town council to destroy their property in the name of urban reform. The owners lost their suit.

15. For details on the circumstances surrounding this view, see Archivo General de Simancas: Consejo Real: legajo 39-III-1, 2.

16. *Os desenhos das antigualhas que vio Francisco d'Ollanda*, ed. E. Tormo (Madrid, 1940), 42r.

17. Pedro de Medina, *Libro de las grandezas y cosas memorables de España* (Sevilla, 1548). The original reads: "podra servir de manual o memoria de las mas señaladas y principales joyas que en esta España tiene."

18. See Botero, *Reason of State.*

19. A revised edition of Medina's book, *Primera y segunda parte de las grandezas y cosas memorables de España* (Alcalá de Henares, 1590), adhered to the same representational scheme, the only significant addition being an illustration of an entire city—walls, plaza, citadel, castle, houses—used interchangeably for capital cities such as Lisbon and Madrid.

20. My understanding of *ad vivum*, a term whose precise meaning is much discussed, derives in part from Claudia Swan, "*Ad Vivum, Naer Het Leven*, from the Life: Defining a Mode of Representation," *Word and Image* 11 (Oct.–Dec. 1995), 353–72.

21. If a "portrait" in the sixteenth century suggested, as the Italian art theorist Vincenzo Danti defined it in 1567, "the portrayal or reproduction of reality as we see it," Hoefnagel's Spanish townscapes appear rather to fall within the category of what Danti defined as "imitation," whose meaning was "to imitate, or to represent reality as it might or ought to be." On these distinctions, see Swan, "*Ad vivum*," 355.

22. Georg Braun, in the introduction of volume 1 of the *Civitates*, offered a different if somewhat far-fetched (and artistically mistaken) interpretation of these and other genre scenes included with some of the views he had published. Responding to critics who feared that the publication of plans and views of Europe's cities would render them vulnerable to Turkish attack, Braun explained that the appearance of human figures in these views would help to deter such an eventuality since the Ottomans, as Muslims, abhorred the representation of the human form.

23. Lucia Nuti, "The Mapped Views by Georg Hoefnagel: The Merchant's Eye, the Humanist's Eye," *Word and Image* 4 (1988): 545–70.

24. For Weiditz's drawings, see Theodor Hampe, ed., *Das Trachtenbuch des Christoph Weidtiz von seinen Reisen nach Spanien (1529)* (Berlin, 1927).

25. For an introduction to the term "curiosity" and its meaning in sixteenth-century Europe, see Gérard Defaux, *Le curieux, le glorieux, et la sagesse du monde dans la premiére moitié du xvie siècle* (Lexington, Ky., 1982). For sixteenth-century books on human diversity, see Jean Ceard, *La nature et les prodiges: L'insolite au xvie siècle, en France* (Geneva, 1977), esp. 252–91. On collecting, see Krzysztoff Pomian, *Collectioneurs, amateurs et curieux: Paris, Venise: xvie-xviie siècle* (Paris, 1990), esp. 61–81. America's role in stimulating interest in the marvelous is examined in Hugh Honour, *The European Vision of America* (Cleveland, 1975) and Joy Kenseth, "The Age of the Marvelous: An Introduction," in Hood Museum of Art, *The Age of the Marvelous* (Lunenburg, Vt., 1991), 25–59.

26. The quotation is from the cartouche accompanying Van den Wyngaerde's 1553 view of Genoa. See Egbert Haverkamp-Begemann, "The Spanish Views of Anton van den Wyngaerde," *Spanish Cities of the Golden Age: The Views of Anton van den Wyngaerde*, ed. Richard L. Kagan (Berkeley, 1989), 54.

27. See Richard L. Kagan, "Philip II and the Art of the Cityscape," *Journal of Interdisciplinary History* 17 (1986): 115–35 (reprinted in *Art and History: Images and Their Meaning*, ed. Robert I. Rotberg and Theodore K. Rabb [Cambridge, 1988], 115–35).

28. On cartographic ties between Spain and the Low Countries in the sixteenth and seventeenth centuries, see the volume *De Mercator a Blaeu: España y la edad de oro de la cartografía en las diecisiete provincias de los Paises Bajos* (Madrid, 1995).

29. The importance Spaniards attached to these artefacts, particularly those of Roman origin, is reflected in Ambrosio de Morales, *Las antigüedades de España* (Alcalá de Henares, 1575), as well as the *antigüedades y grandezas* genre of local Spanish history. These books paid particular attention to ancient coins, lapidary inscriptions, statues, and other artefacts in order to draw attention to a city's importance in Greek and Roman times. An early example, dating from 1579, is Diego de Villalta, "Historia de la antigüedad de la Peña de Martos," ed. Joaquín Codes y Contreras (Madrid, 1923).

30. A contemporary discussion of the extent to which urban grandeur rested on commerce and agriculture may be found in Botero, *Reason of State.*

31. For Cuelbis, see British Library, mss. Harley 3822, "Thesoro chorographico de las espannas por el señor Diego

Cuelbis." This travel account includes over two dozen small townscapes, the majority of which are little more than copies of prospects previously published in the *Civitates Orbis Terrarum*. Most of the cities depicted in this volume are generally illustrated twice, first in a pen-and-brown-ink drawing—presumably by Cuelbis himself—and then in an engraved copy by a certain Francesco Valeço or Vallegio. The two versions are not always identical. Vallegio added figures and other folkloric scenes, most of which were copied from Hoefnagel's views.

32. Completed in 1593, the original manuscript is conserved in the University of Coimbra library. See Fernando Gabriel Martín Rodríguez, *La primera imagen de Canarias: Los dibujos de Leonardo Torriani* (Santa Cruz de Tenerife, Colegio Oficial de Arquitectos de Canarias, 1986).

33. Torriani's views are comparable to the series of *plattengronden* of the Netherlandish cities executed, again for Philip II and primarily for military purposes, by Jacques van Deventer in the 1550s and 1560s. For more on this project, see Kagan, "Philip II and the Art of the Cityscape," 120, and Geoffrey Parker, "Maps and Ministers: The Spanish Habsburgs," in *Monarchs, Ministers, and Maps,* ed. David Buisseret (Chicago, 1992), 124–52.

Philip II was also responsible for the "Descripción de las marinas de todo el reino de Sicilia," an atlas of Sicily by Tiburzio Spannochi that includes views of that island's coastal cities. Conserved in the Biblioteca Nacional, Madrid, ms. 788, this important manuscript remains unpublished.

34. For more on this plan, see Fernando Benito Doménech, "Un plano axonométrico de Valencia diseñado por Manceli en 1608," *Ars Longa: Cuadernos de Arte* [Valencia] 3 (1992): 29–37.

35. Quoted in Swan, *"Ad vivum,"* 353. Kepler used this particular expression in a conversation with Sir Henry Wooten in 1620 and with particular reference to a landscape he had recently completed.

36. These and other plans of Madrid are reproduced in M. Molina Campuzano, *Planos de Madrid en los siglos xvii y xviii* (Madrid, 1960). For Teixeira's *topografia,* see Jesus R. Escobar, "The Plaza Mayor of Madrid: Architecture, Urbanism, and The Imperial Capital, 1560–1640," Ph.D dissertation, Princeton University, 1996, pp 257–59.

37. On the Nicolay map, see the insightful discussion by Hilary Ballon, *The Paris of Henri IV* (Cambridge, Mass., 1991), 212–47.

38. Suggestions that Vermeer made use of a *camera obscura* for his *View of Delft* remain inconclusive. See the catalog of the exhibition, *Johannes Vermeer* (National Gallery of Art, Washington D.C., 1996), 25–27, 69–74.

39. See R. M. Vorstman, "Schilderijen, prenten en tekeningen: De haven van Santander," *Jaarverslag: Vereeniging Nederlandsch Historisch Scheepvaart Museum,* Amsterdam (1983), 8–9. José Luis Casado Soto, *Santander, una villa marinera en el siglo xvi* (Santander, 1990) mistakenly attributes the drawing to Hoefnagel.

40. For Baldi, see J. Maglialotti, *Viaje de Cosimo de Médicis por España y Portugal (1668–1669),* ed. A. Sánchez Rivero y A. Mariutti (Madrid, n.d.)

41. For more on this painting, see Elizabeth Trapier, "Martínez de Mazo as Landscape Artist," *Gazette de Beaux Arts* (May, 1963): 293–310.

42. Antonio Palomino, *Vidas,* ed. Nina Ayala Mallory (Madrid, 1986), 223. Palomino's comments, first published in 1724, were meant also to apply to Mazo's painting, *The Entrance of Philip IV into Pamplona,* a copy of which is now in London's Wellington Museum.

43. Van den Wyngaerde's view of Zaragoza (1563) is reproduced in Kagan, ed., *Spanish Cities,* 147.

44. For more on the history of view making in early modern Spain, see Richard L. Kagan and Fernando Marías, *Urban Images of the Hispanic World, 1500–1750* (Yale University Press, forthcoming). For an introduction to maps and views of Spanish cities in the eighteenth century, see Carlos Sambriucio, *Territorio y ciudad en la España de la Ilustración,* 2 vols. (Madrid, 1991).

45. Archivo Histórico Nacional: Códices 1368b, Privilegia per Serenissimos reges civitati, Oriole concessa, fol. 46. This manuscript, a copy of a cartulary dating from 1406, is briefly discussed in Justo García Morales, "Vista del castillo y villa de Orihuela: Cartulario de Orihuela: AHN Códices 1368," *Fiestas Moros y Cristianos* (Orihuela), *año* 1981 (no pagination).

46. See Rafael Martín de Viciana, *Tercera parte de la crónica de la inclita y coronada ciudad de Valencia y su reino* (Valencia, 1563; fascimile ed., Valencia, 1980), 321–331.

47. Luis de Toro, "Palentiae urbis et eiudem episcopotus descriptio," Biblioteca Universidad de Salamanca, ms. 2650. See also Luis de Toro, *Descripción de la ciudad y obispado de Plasencia,* ed. and trans. Marceliano Sayáns Castaños (Plasencia, 1961).

48. For the abundance of views in one German city, see Jeffrey Chipps Smith, "Renaissance Nuremberg as the Ideal City: Observations on the Art of Civic Imaging and Political Control," (n.d.). Turin offers another example of a city well served by plans and views of various sorts. See Martha Pol-

lak, *Turin, 1564–1680* (Chicago, 1991). For the full collection of Italian topographical views and plans, see *La città nella storia d'Italia,* ed. Cesare De Seta (Bari, 1979–).

49. For the concept of the "imageability" contained in certain physical objects, see Lynch, *Image of the City,* 9.

50. *Historia del arte valenciana,* ed. Vicente Aguilera Cerni (Valencia, 1986), 317.

51. For these views of Seville, see *Iconografía de Sevilla, Tomo I: 1400–1650,* ed. María Dolores Carbrera Laredo (Madrid, 1988).

52. One exception is the pen-and-ink prospect included in Juan de Mal-Lara, *Recibimiento que hizó la ciudad de Sevilla a don Phelipe* (Seville, 1570). Even so, an outsized giralda dominates the view.

53. For this incident, see Diego Ortiz de Zúñiga, *Anales eclesiásticos y seglares de . . . Sevilla* (Seville, 1795), 2:252–53.

54. For representations of Mary in sixteenth- and seventeenth-century Spain, see Suzanne Stratton, *The Immaculate Conception in Spanish Art* (Cambridge, 1994).

55. Zurbarán's *Immaculate Conception* (1630), now in the Museo del Diocesís de Sigüenza and reproduced in Julián Gallego and José Gudiol, *Zurbarán, 1598–1664* (New York, 1977), 154.

56. For more on this painting, see Jonathan Brown and Richard L. Kagan, "El Greco's *View of Toledo,*" in *Figures of Thought: El Greco as Interpreter of History, Tradition, and Ideas,* ed. Jonathan Brown (Washington, D.C.: National Gallery of Art, *Studies in the History of Art,* vol. 11, 1982): 19–30.

57. Mauro Castella Ferrer, *Historia del apostol de Jesus Christo Santiago Zebedeo Patron y Capitan de las Españas* (Madrid, 1618).

58. See Jose Martín Gómez-Moreno Calera, *El arquitecto granadino Ambrosio de Vico* (Granada, 1992).

59. For Castro's involvement in the Christianization of Granada's past, see Miguel José Hagerty, *Los libros plúmbeos de Sacromonte* (Madrid, 1980), and, more recently, Julio Caro Baroja, *Las falsificaciones de la historia* (Barcelona, 1992), 118–43.

60. Prior to their expulsion from the city in 1570, the Albaicín was home to most of Granada's *moriscos,* or converted Moors. For a summary of urban reforms in sixteenth-century Granada, see Antonio Luis Cortés Peña y Bernard Vincent, *Historia de Granada* (Granada, 1986), 3:27–43.

61. My argument here follows that of Antonio Moreno Carrido et. al., "La Pla [n] taforma de Ambrosio de Vico: Chronología y Gestación," in *Arquitectura de Andalucia oriental* 2 (1984):6–12, and José Luis Orozco Pardo, *Christianópolis: urbanismo y contrareforma en la Granada del seiscientos* (Granada, 1985).

62. See Quesada, *La idea de ciudad,* 41–57.

63. For more on the "false chronicles," see Caro Baroja, *Las falsificaciones,* 163–187. As recently demonstrated by Fernando Checa, *Felipe II: Mecenas de las Artes* (Madrid, 1992), it was Philip II, starting in the 1560s, who initiated the recovery of Spain's "Christian antiquity."

64. For these histories, see my "Clio and the Crown: The Writing of History in Habsburg Spain," in *Spain, Europe, and the Atlantic World,* ed. Richard L. Kagan and Geoffrey Parker (Cambridge, 1995), 73–99.

65. Of particular interest is the series of paintings devoted to the life of Saint Julián which is still to be found in Cuenca's cathedral. See Fernando Benito Domenech, "Una singular serie de cuadros sobre la vida de San Julián," *Archivo de arte valenciano* 50 (1979):70–75. The lost view of Cuenca, attributed to Cristóbal García Salmeron (1603–66), a native *conquense,* was described in 1668 as "un lienzo de dos varas y media de largo y dos de alto de la ciudad de Cuenca y sus fiestas que hicieron el año 1642 al Rey nuestro señor que está en gloria [Philip IV], marco negro: original de Xpoval García." See Diego Angulo Iñiguez and Alonso E. Pérez Sanchez, *Pintura toledana: Primera mitad del siglo xvii* (Madrid, 1972), 370.

66. The painting is documented in Diego Angulo Iñiguez and Alfonso E. Pérez Sanchez, *Pintura madrileña: Primer tercio del siglo xvii* (Madrid, 1969), 249.

67. For this and other seventeenth-century views of Barcelona, see *Retrat de Barcelona* (Barcelona, 1995), 1:85–110.

68. "Descripcion del Reino y Obispado de Jaen," Biblioteca Nacional, ms. 1180.

69. For more on these paintings, see Jesús Urrea, "Tres vistas de Valladolid en el siglo xvii," *Boletín de la Real Academia de Bellas Artes de la Purísima Concepción de Valladolid* 29 (1994):197–208. This series of paintings illustrating a community united by faith anticipated Felipe Gil de Mena's painting of the auto-da-fé staged in Valladolid's *plaza mayor* in 1667 as well as Francesco Rizi's later rendition of a similar event held in the *plaza mayor* of Madrid.

70. For these images, see Juan Miguel Serrera y Alberto Oliver, *Iconografía de Sevilla, 1650–1790* (Madrid, 1989).

71. See Pierre Lavedan, *Représentation des villes dans l'art du moyen âge* (Paris, 1954), and Jürgen Schulz, "Jacopo de' Barbari's View of Venice: Mapmaking, City Views and Moralized Geography before the Year 1500," *Art Bulletin* 60 (1978):425–72.

72. Little is known about Garau except that he was a na-

tive of Palma de Mallorca and the author of a treatise on the efficacy of prayer prior to his death in 1657. For more information, see Diego Zaforteza y Musoles, *La ciudad de Mallorca: ensayo histórico-toponímico* (Palma de Mallorca, 1987).

73. See Parker, "Maps and Ministers," 145.

74. For Spanish engineers, see Nicolas García Tapia, *Ingenería y arquitectura en el renacimiento español* (Valladolid, 1990), 60–67. Although Spaniards only accounted for about one third of the military engineers in the crown's employ, over 70 percent of a total of 168 *ingenieros* known to have worked in sixteenth-century Spain were native-born. For the detailed study of one such individual, see Angel Laso Ballesteros, "Tradición y necesidad: La cultura de los inginieros militares en el siglo de oro: la biblioteca y la galería del capitán don Jerónimo de Soto," *Cuadernos de historia moderna* 12 (1991):83–109.

75. For contemporary comments about the shortage of detailed "maps" and "descriptions" in seventeenth-century Spain, see Fernando Bouza, "Cultura de lo geográfico y usos de la geografía entre España y los Paises Bajos durante los siglos xvi y xvii," in *De Mercator a Blaeu,* 62–63. Particularly noteworthy are the comments of the Flemish cartographer, Jean Charles della Faille, who was responsible for maps needed for a military campaign against the Portugal in 1641. Referring to the area around the border town of Ciudad Rodrigo, della Faille lamented that "this entire region is not very well known; there are no maps, let alone any descriptions."

FOUR

Military Architecture and Cartography in the Design of the Early Modern City

Martha Pollak

The Ubiquity of War

"The most beautiful aspect of Architecture is surely that which deals with cities," wrote Pietro Cataneo in the dedication of his *Four First Books on Architecture* to Enea Piccolomini, a descendant of Pope Pius II who had been such a great city builder and lover of architecture. "But," he continued, "since cities are now threatened by artillery which the ancients did not possess, I will demonstrate how to build cities differently from theirs so as to defend them from a menace that was previously unknown" (figure 4.1).[1] Since the association between architecture and the city had been made earlier by Leon Battista Alberti, Filarete (Antonio Averulino), and Francesco di Giorgio Martini,[2] the novelty of Cataneo's idea consisted in the further association of city planning with military architecture. But by defining in print the defense of the city as the domain of the architect, he merely made a formal claim for a practice already accepted two generations earlier. Thus when in 1482 Leonardo da Vinci offered his services to Ludovico Sforza, the duke of Milan, he recommended his own skills in a ten-point letter where his acknowledged artistic achievements in painting, sculpture, and architecture are relegated to the last point, using the first nine to describe enticingly for the duke the war machines and ferocious firearms that he had invented. He commended his own abilities in war, claiming that he could destroy "every 'rocca' or other fortress."[3] When Leonardo, with Francesco di Giorgio Martini and Alberti, claimed military architecture for the Renaissance architect and artist, he bound together two seemingly irreconcilable activities: construction and destruction. Furthermore, he associated these activities with that of surveyor and cartographer, and he may have been the first artist and military architect to produce an accurately measured and drawn ichnographic plan of a city—that of Imola (figure 2.27 above)—which became a fundamental cartographic document for the subsequent representation of the European city.[4]

These same, early Renaissance architectural theorists concerned themselves intensely with the conception and planning of the ideal city. Francesco di Giorgio Martini devoted some of his most elaborate drawings to this problem, while Filarete and Leonardo da Vinci not only provided designs but added their own conceptions of social order which were expressed through the architectural form and the general layout of their proposed towns. Based on aesthetic principles simultaneously developed in painting and in architecture, their conceptions were centralized plans with radial or orthogonal grid streets, centrally located piazzas, and gates and bastions on

PRIMO. 20

recinto della città guidati. & de i membri dentro la muraglia, per eſſere coſi piccoli i diſegni moſtri e da moſtrarſi, non ſi ſon fatti nell'alzato di loro proſpettiue ſe non il ter raglio. ne di quelli ancora non ſi ueggono le porte: perche non ſi dimoſtrarebbono di alcuna apparenza.

FIGURE 4.1. Pietro Cataneo, Plan of fortified city with pentagonal fortress. Woodcut. From *I quattro primi libri di architettura*, Venice, 1554, fol. 20. Photo courtesy of The Newberry Library, Chicago.

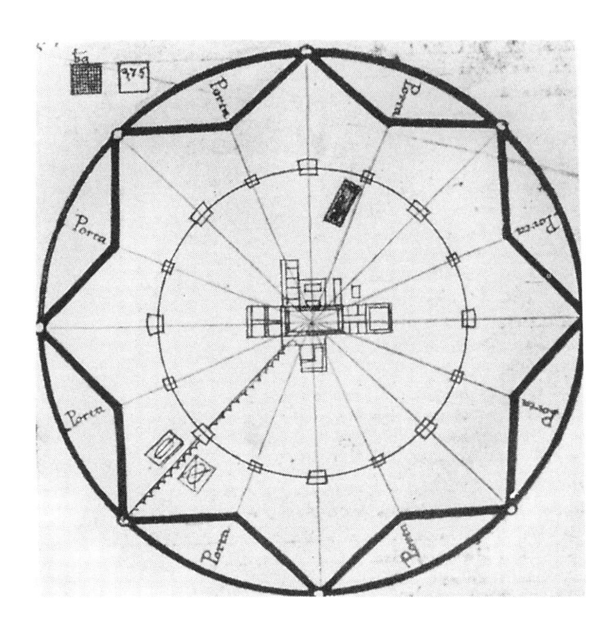

FIGURE 4.2. Filarete (Antonio Averulino), Plan of the ideal town of Sforzinda. Pen and ink, National Library, Florence.

axes with one another. Their compositions were based upon symmetry, hierarchy and harmony of the individual parts with one another.

Leonardo's sketches of Milan show familiarity with Filarete's suggestions, made while he was employed at the Milan court (figures 2.13 above, 4.2). Filarete's radially laid out ideal city plan was in distinct contrast to available models such as the castrum plan of Roman cities. A good example of the Roman model is the plan of Turin, but Albrecht Dürer's conception of the same castrum is as impressive (figures 4.3, 4.4). The radial and the orthogonal city plans became the two distinct models for urban design for the next three centuries.

The implicit control suggested by the rigorous geometry suited not only the sixteenth-century dictatorships but also, perhaps even more, the absolute monarchies of the seventeenth century. It has been said that the ideal city of the fifteenth- and sixteenth-century utopian architects and philosophers—men such as Leonardo, Francesco di Giorgio Martini, and Erasmus—became the ideal fortress of the seventeenth-century military architects.[5]

Leonardo's attitude, as seen in his application to the duke of Milan, demonstrated the centrality of war in the development of the early modern state and of absolute monarchy, the form of government refined in the seventeenth century. War radically transformed the appearance and functions of the early modern European city.

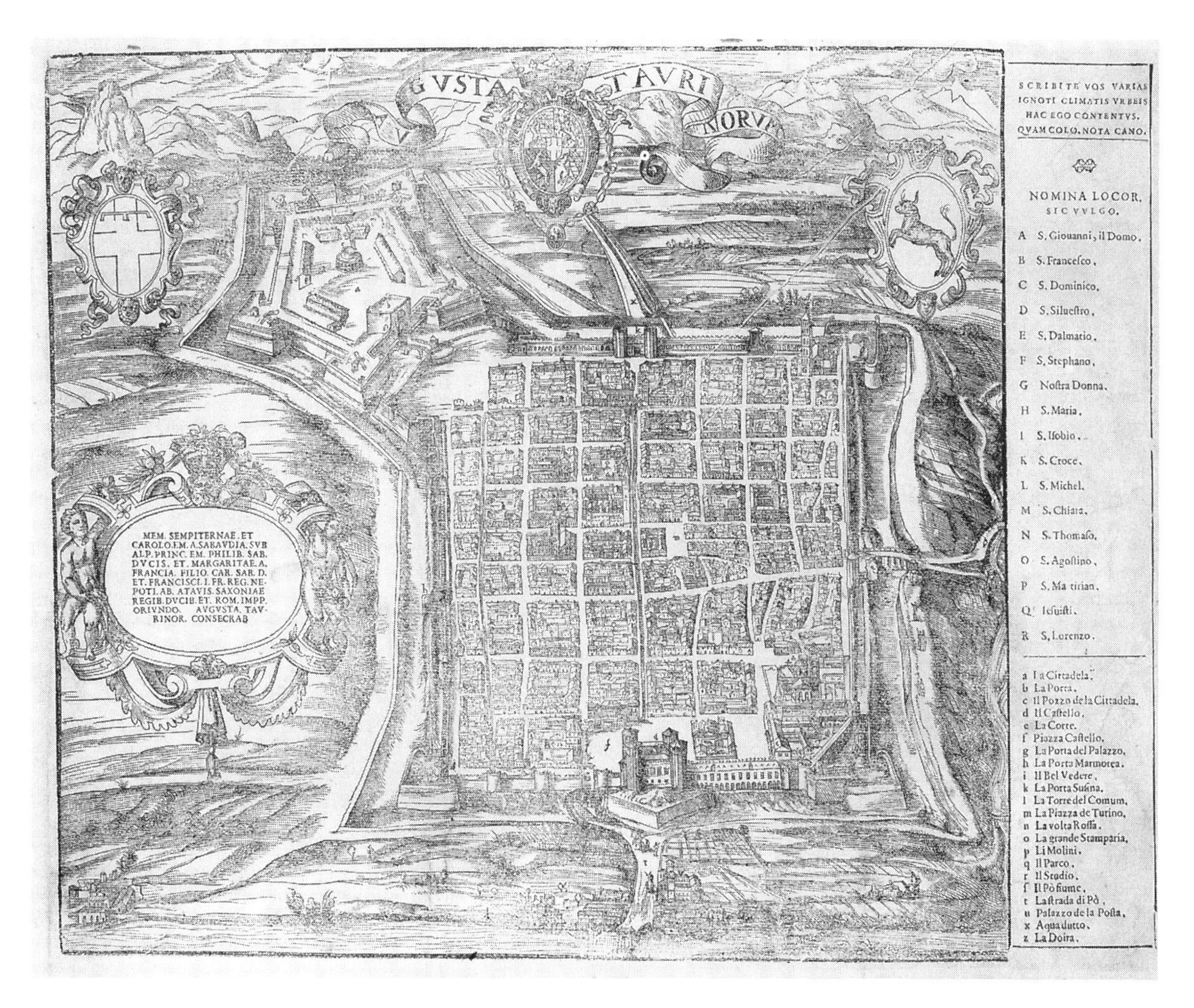

FIGURE 4.3. Giovanni Criegher and Giovanni Caracha, Bird's-eye view of Turin. Woodcut, 1572. Biblioteca Reale, Turin.

Most conflicts were resolved through sieges of towns, since the question was no longer to establish the greater ability and strength among equals (as in the morally defensible clashes inherited from the Middle Ages) but to demonstrate sovereignty.[6] Consequently, the great military conflicts of the sixteenth and seventeenth centuries were known by the names of the cities around which they were contested: Casale, Breda, La Rochelle are among them (figures 4.5a, 4.5b).[7] Fortification of towns became paramount then for the defense of a sovereign's territory and claim to nationhood. Parallel with the interest and need for more efficient, resistant, and affordable fortification, there developed a field of research in military strategy. Thus the practical requirements of war preparation were matched by an equivalent theoretical counter-

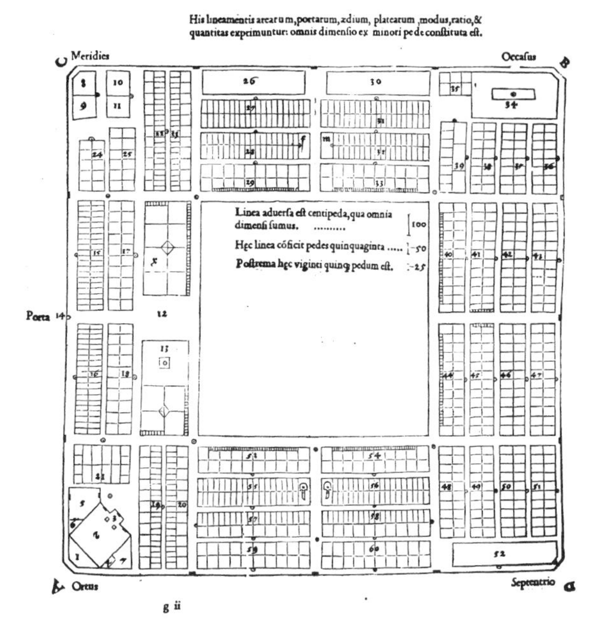

FIGURE 4.4. Albrecht Dürer, Plan of fortified city. Woodcut. From *Etliche Underricht zu Befestigung der Stett, Schloss, und Flecken,* Nuremberg, 1527. Photo courtesy of The Newberry Library, Chicago.

part focused on the art, philosophy, and science of war.

The Writers of Treatises on Military Architecture

THE EXTENSIVE MILITARY literature of the early modern period illustrates vividly the interest taken in matters associated with war, and a significant segment of this literary output is focused on the problems of military architecture—the permanent fortification of cities, towns, and strategically located sites. While some early texts on war were authored by nonspecialist writers, such as the protofeminist author Christine de Pisan, the papal abbreviator Roberto Valturio, and the political philosopher Niccolò Machiavelli,[8] the earliest innovations in Renaissance military architecture are owed to artists and architects such as Francesco di Giorgio Martini, whose architectural treatise remained unpublished until the nineteenth century, and Leonardo da Vinci.[9] Early fortifications were powerfully visualized by Michelangelo and by the architects of the Sangallo family, whose observations and drawings as well as realized fortifications were crucial in the development of an aesthetic of military architecture.[10] In this the polygonal bastion was the essential feature.[11]

The bastioned trace was designed to be beautiful by its fifteenth- and sixteenth-century architects and was expected to be awesome by its commissioners.[12] Thus Montaigne, despite the great rain and his eagerness to move on, was touched by the terrible beauty and great size of Milan's Castello, "une edifice très grand et admirablement fortifié," whose perimeter he took in entirely; Turin's fortifications, reputed as among the most powerful in Italy by 1630, were described as *bellissime* by awed visitors, while Louis XIV did not mind the high cost of fortification as long as the result was a *belle place.*[13]

The shift of emphasis from an artistic to a scientific approach is chronologically framed by Albrecht Dürer's treatise, published in 1527 and thus the earliest illustrated book on military architecture, produced by a Renaissance artist known as one of the foremost engravers of Western art, and Vauban's treatise, the work of an influential field marshal and strategist of Louis XIV.[14] While artists and architects dominate military architecture during the sixteenth century, military theorists in the seventeenth century pursued a wide range of artistic, literary, scientific, and editorial activities, as they studied war for itself and as a fashionable subject. Their ranks included mathematicians such as Andreas Cellarius, Niccolò Tartaglia, Samuel Marolois, and Simon Stevin; cartographers, printers, and publishers such as Henry Hondius, Jean Dubreil, Ni-

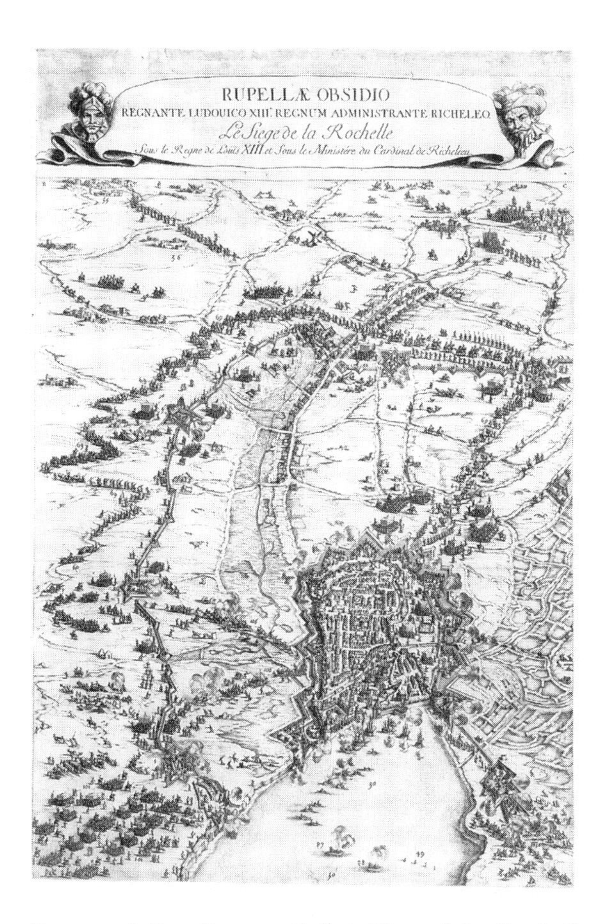

FIGURE 4.5a. Jacques Callot, Plan of La Rochelle under siege, 1630. Copperplate engraving. Photo courtesy of The Newberry Library, Chicago.

colas de Fer, and Daniel Bodenehr; Jesuits involved in teaching such as Georges Fournier, Andrew Tacquett, and Donato Rossetti; historians like Girolomo Maggi, Galeazzo Gualdo-Priorato, Jacques Ozanam, and L. C. Sturm; while Coehoorn and Vauban were the busiest and most innovative strategists. But most of the authors were artillery officers, such as Galasso Alghisi, Gabrio Busca, Pietro Sardi, or engineers, such as Daniel Speckle, Jacques Perret, Adam Fritach, and Nicolas Goldmann, whose writings

FIGURE 4.5b. Jacques Callot, Plan of La Rochelle under siege, 1630. Copperplate engraving. Photo courtesy of The Newberry Library, Chicago.

were largely based on their personal experience in war and in architecture.[15]

They dedicated these publications to the most powerful members of the aristocratic elite. In his study of some of these treatises Sir John Hale has convincingly demonstrated, through an analysis of their title pages, that the dedications—supported by the composition of engraved frontispieces—attempt to uplift the social and intellectual standing of the author by association with a famed military leader or diplomat.[16] Thus the production of military treatises by the practitioners of military architecture can be seen as an effort not only to endow the profession with a theory but also to ennoble it by giving it a measure of responsibility for cultural and political life. The attempt was to endow a practical profession with an intellectual context and, by associating it with the interests of the powerful, to elevate the profession's social standing as well.

"Space as analogous with geometry" has been a way of interpreting the seventeenth-century mental attitude that extended naturally into military architectural theory.[17] Conception of nature as either God-given geometrical order and mathematical reality or as a wilderness, a labyrinth to be navigated through the use of method, encouraged scientific and theoretical research. For Descartes, everything seemed to be a science, his universe a "grid of arithmetical relations."[18]

The definitive welding of military architecture with mathematics and geometry occurred through the scientific approach adopted in the seventeenth century. While Buonaiuto Lorini in his *Delle fortificationi* (Venice, 1596 and 1609) guided military engineering to the threshold of modern science, Jean Errard transformed the mechanical art of fortification into a scientific and therefore "perfectly demonstrable art" in his *La Fortification reduicte en art et demonstrée* (Frankfurt, 1604).[19] Perret identified the science of fortification with geometric construction in his *Des Fortifications et artifices* (Paris, c. 1601), and Freitag's treatise *Architectura nova et aucta, oder newe vermehrte Fortification* (Leiden, 1642) is one in a long chain of works on geometry and trigonometry in which the author believes that everything, even the most pragmatic aspects of fortification, can be calculated in advance.[20] This notion of accountability was supported in Marolois's *Fortification ou architecture militaire tant offensive que defensive* (The Hague, 1615); he believed in the dominance of theoretical design amply realized in Vauban's *cahiers de charge*.[21]

Some theorists avowed preferences for certain regular polygons. The "Dutch Vauban" Menno, baron of Coehoorn, writes at length in his *Nieuwe Festingbow* (Leeuwarden, 1685) about the "royal hexagon," which is also Sardi's favored fortification form, illustrated in his *Corona imperiale della architettura militare* (Venice, 1618). Errard's book on fortification begins with the design of the hexagon (according to him the first regular polygon that can be easily fortified), but then goes on to demonstrate the composition of polygons with up to twenty-four sides. This obsessiveness is implicitly criticized by Matthias Dögen, who stops his own demonstrations in *Architectura militaris moderna* (Amsterdam, 1647) at the octagonal fortress, pointing out that no one was building greater regular fortifications (the nine-sided Palmanova near Venice had been the only exception).[22] Subsequently, B.F. Pagan claimed in *Les Fortifications* (Paris, 1645) that his own pentagonal and dodecagonal fortifications were the most powerful, while Antoine de Ville—his most talented competitor—devoted many pages of *Fortifications ou l'ingenieur parfait* (Lyon, 1629) to the analysis of actual pentagonal fortresses, in Turin (see figure 4.3), Antwerp, Casale Monferrato, and Rome.

Indeed, the pentagon seems to have endlessly fascinated military architects and theorists; the form, there from the beginning of cannon fortification, remained meaningful throughout the sixteenth and seventeenth centuries and still has powerful military associations today. The pentagon appears to have been the most popular form

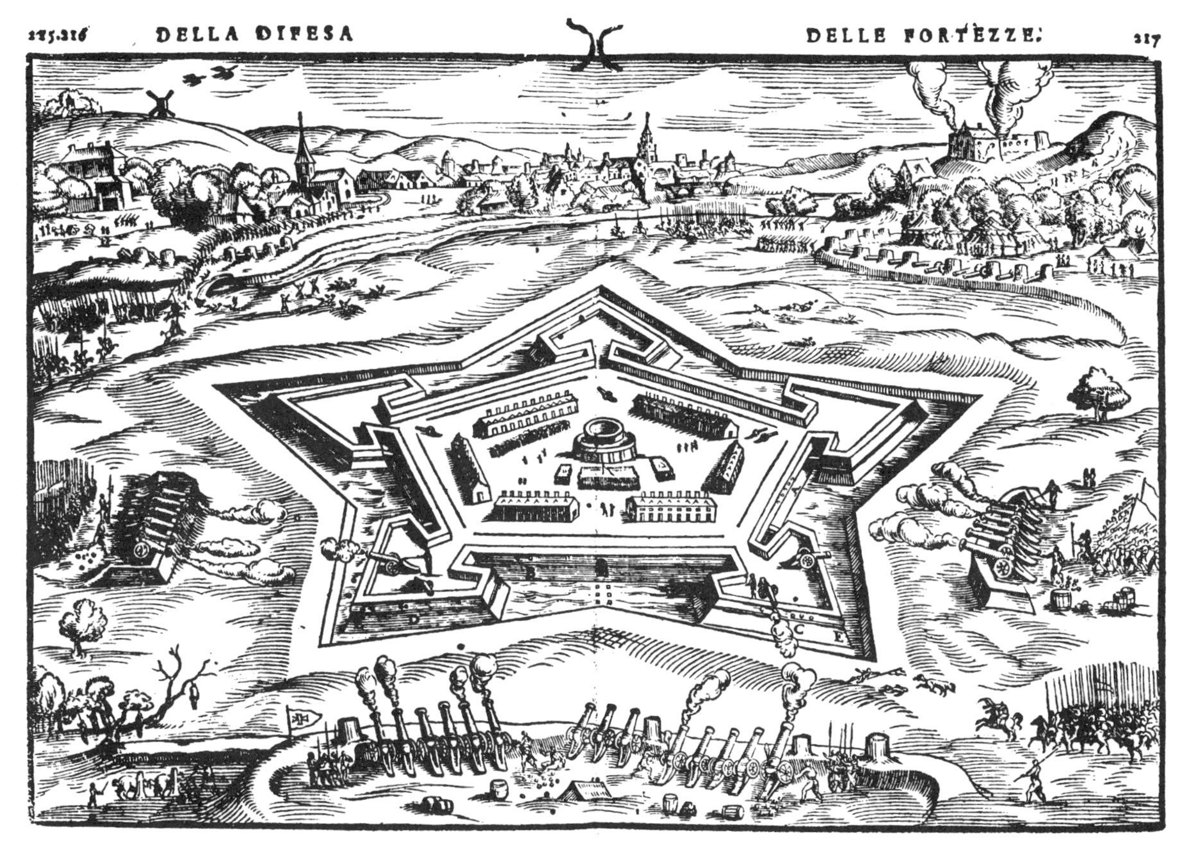

FIGURE 4.6. Gabrio Busca, View of pentagonal fortress. Woodcut. From *Della espugnatione et difesa delle fortezze libri tre*, Turin, 1585. Photo courtesy of The Newberry Library, Chicago.

for citadels and fortresses (figure 4.6); while the latter can be interpreted as cities in microcosm, the pentagon itself is the sign of the Microcosm, held in occultism to be the most powerful means of conjunction. Used by all secret and occult societies, it seems an appropriate form to manifest the desires of the secretive military architects. According to the Pythagoreans, five represents fire; according to Cornelius Agrippa (1533), five exorcises evil and is the number of justice, and the five-pointed star is associated with the pentagram. Symbolizing man after the fall, it is associated with life here on earth. Esoteric thinking relates the number five to man's five extremities. Association of five with the human body is common in the Renaissance, but it is also found all over the world, from England to the Far East.

Although there is no direct evidence of specific occult practices among sixteenth- and seventeenth-century military architects, they would nevertheless have been aware of the significance of the numbers they were working with as numerology and cabalistic studies enjoyed great popularity during this period.[23] Their own insistence on the use of specific numbers and the fortifications conceived around them suggests that they may have been attempting to accommodate numerology and cabalism in military studies.

The association with the human body is fundamental, of course, and memorably illustrated in Leonardo's influential placement of the human figure within a circle and a square. The importance of conjunction is evident since the citadel and the city were to be indelibly linked. This conjunction was not seen necessarily as a positive connection: in Machiavelli's writing the citadel is considered an evil subjugation of the city. In this context the form of the conjunction might be interesting since the placement of the five-pointed polygon has various meanings. With the one point in the ascendant it was read as the sign of Christ; with two points in the ascendant it was the sign of Satan. The pentagonal fortresses designed by Francesco de' Marchi, and those of Rome, Florence, Turin, Antwerp, Parma, and Piacenza orient two bastions toward the city and three toward the surrounding countryside.[24] The fortress is always at the edge of the city, interwoven with the latter's fortifications (figure 4.7).[25]

But the fortress does not simply defend the city. The citadels were built in order to subdue the local population, to act as a deterrent to popular uprisings, and as ultimate shelter for the sovereign ruler. In order to take the citadel, the city itself first had to be taken by a besieging army. But what can we make of the actual design of the connection between the fortifications and the citadel? The practical reason given for training two bastions against the city and three against the surrounding countryside is that the enemy from outside is nonetheless more dangerous than the enemy within. But the orientation of the pentagon with two points toward the city and one toward the countryside implies powerful occult readings. With God at the single top point, the two points most distant from it, separated from the peak by intellect and civilization and highest matter, signify lowest matter and moral corruption. These would be, of course, the attributed causes of urban conflicts such as civil wars which the pentagonal citadel was expected to prevent, or at least to redress.

Representation

But why link military architecture and cartography?

The most influential subject of these treatises for urban history is the representation of fortifications and the cities they surround. While there are suggestions in the sixteenth-century texts that a drawing of a fortress might clarify in advance the problem implicit in a design solution,[26] there seems to have been no systematic method for drawing parts or entire fortifications until the seventeenth century when representation became closely linked to an understanding of the fortress's parts.

The interest in and the need for representation linked military architecture to civil architecture. The use of plan, section, and elevation had been promoted in the sixteenth century by Raphael for the historical documentation of Rome's quickly disappearing ancient monuments, and by Daniele Barbaro, in his edition of Vitruvius's *De Architectura* (1556).[27] The three means of representation, *orthographia, ichnographia,* and *scaenographia,* were first outlined by Vitruvius, but Barbaro substituted *sciagraphia* for the third term, rendered in Italian "profilo" and thus derived the plan, elevation, and profile. Significantly, this system of representation was not consistently applied either in the design or documentary stages of a building until its espousal in these treatises on military architecture.

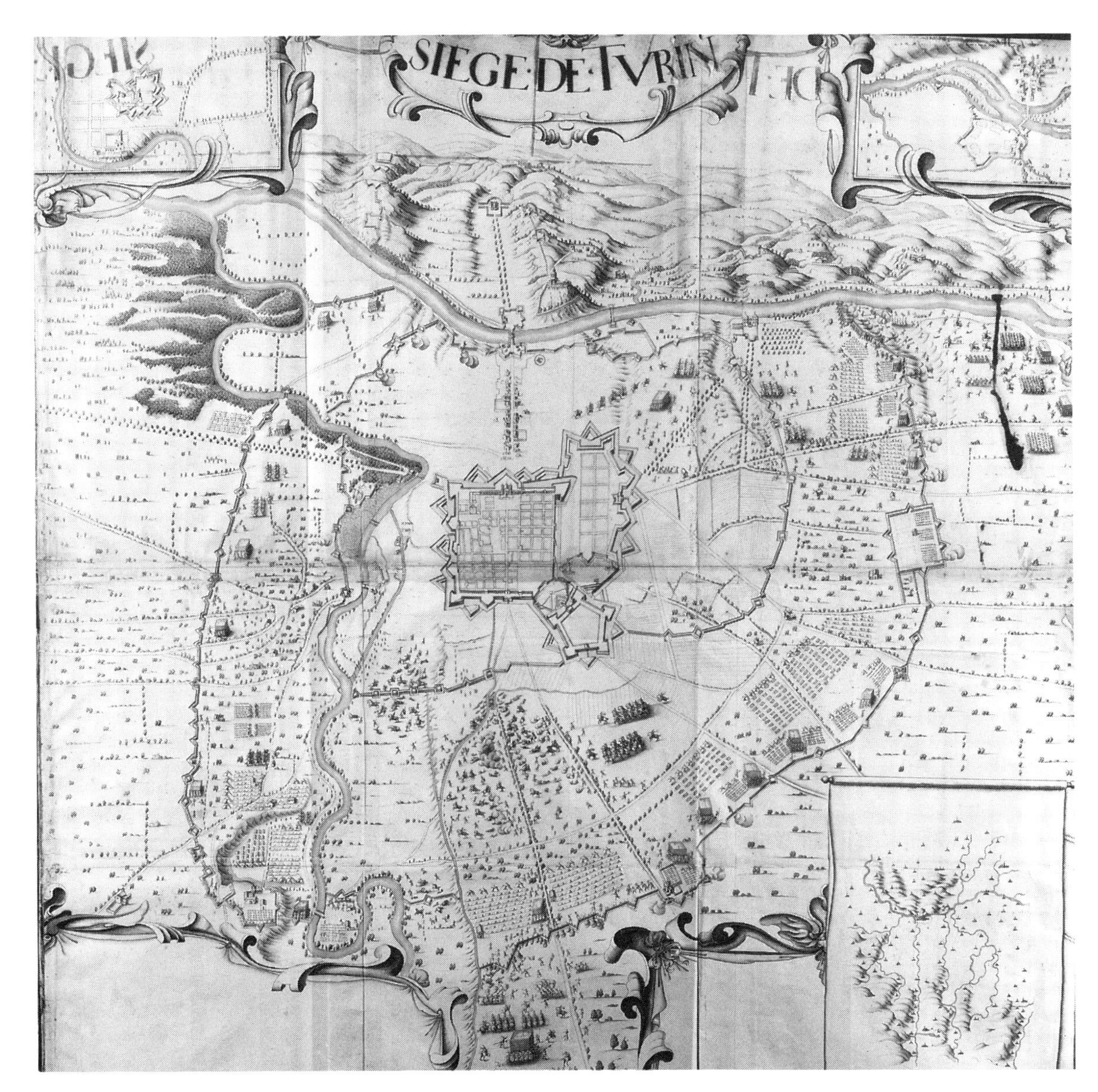

FIGURE 4.7. Giovenale Boetto, View of the siege of Turin, 1642. Copperplate engraving. Cabinet des Estampes, Bibliothèque Nationale, Paris.

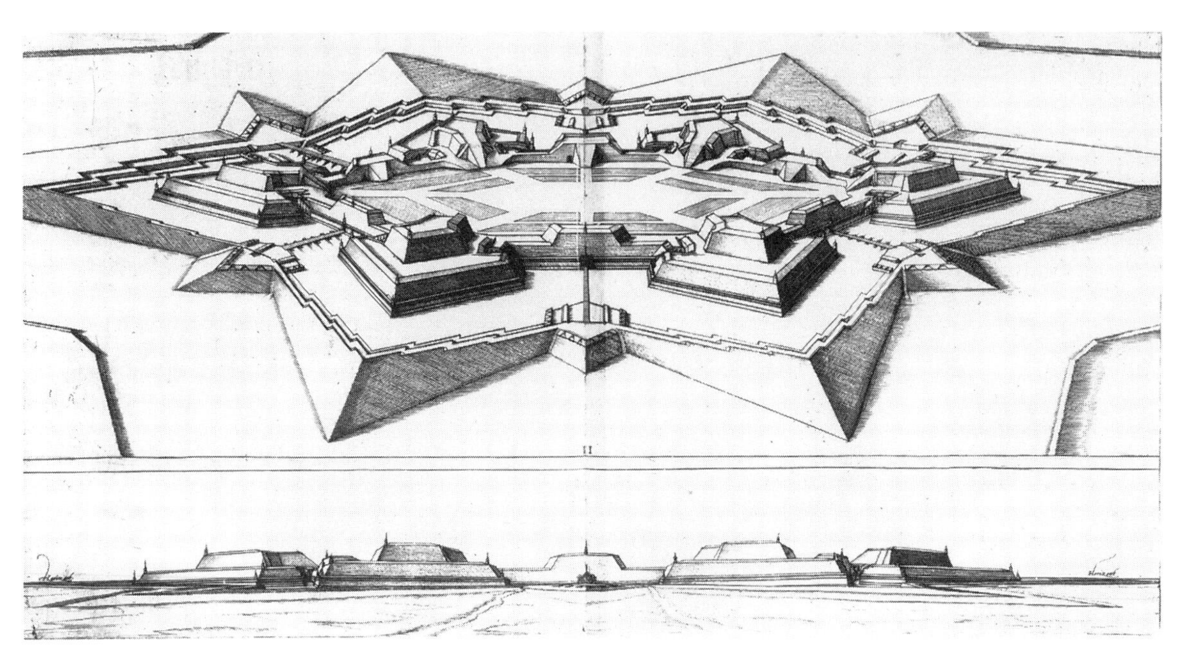

FIGURE 4.8. Daniel Speckle, Bird's-eye view and elevation of a fortress. Woodcut. From *Architectura von Vestungen*, Strasbourg, 1589. Photo courtesy of The Newberry Library, Chicago.

Galileo had also defined the plan and section at the end of the sixteenth century;[28] in 1625 Hondius listed three means of representation—ichnographic, orthographic, and scenographic—but it was Dögen (1648) who convincingly defined the three means of graphic representation that corresponded to these terms: the horizontal section or plan, the vertical section or profile, and the perspective, axonometric, or bird's-eye view; he used these means of representation to explain the terminology of military architecture. This scientific definition of plan, section, and perspective view was of fundamental importance for the ability to visualize fortifications before building them in order to check dimensions, accuracy, and strategical planning, as well as to present a proposed design to a patron. The plan and section provided in addition a very accurate method of documenting existing fortifications and aided in reconstruction and restoration efforts necessitated by constant war.

Significantly, military architecture and its literature altered the image of the city beginning with the earliest published treatises. Through these works military theorists disseminated the image of the European city with a geometrically perfect layout surrounded and defined by impeccably designed fortified defenses. This sharp definition endowed the representations of fortifications with an immediacy that was palpably stronger than the realities of the actual city, whose walls could never be perceived as a whole except in a cartographic illustration. The visual strength achieved through the manipulation of perspective made the fortifications seem even more fearful than they were in actuality (figure 4.8).

Besides adopting and refining methods of representation borrowed from civil architecture, the illustrations of the military treatises benefited from the results of accurate surveying, an endeavor that was in turn greatly favored by the intensive activity of military architects and engineers. The construction of new fortifications and the restoration of old ones, as well as the siege of a town, required accurate topographical site plans. Thus the science of surveying reached a higher degree of perfection through extensive experiment as well as through the instruments invented by some of the same military architects.

Aided by their advanced scientific texts on geometry and driven by a need for secrecy, military architects tended to choose the representation in plan, that is, the horizontal section. This is the most abstract representation of a city and, consequently, the most difficult for an outsider to read. The plan became, with the section or profile, the preferred representational manner for military as well as civic architects; and it became the shorthand for professional drawing, because of its both secretive and abstract qualities. The abstraction was the result of the use of accurate scales, also taught by military theorists, that placed each part of a building, fortification, or city plan in precise relationship to the others, no longer privileging or neglecting parts of the object represented. The proportional plan provided information about the parts of the city which, through the use of scale, could be accurately measured off. The advantage of the abstracted plan was also that it allowed a total and unprejudiced view of the entire fortification and the city it enclosed, whereas the view always emphasized the objects in the foreground.

These improvements in surveying and drafting techniques and the shift in mental attitudes implicit in the adoption of the horizontal and vertical sectional drawings served to improve the art of cartography. Only then could the histories of military architecture, published toward the end of the seventeenth century, become an independent genre as cartographic albums. From its emphasis on the interpretive view, visually powerful but not always accurate, military architecture shifted to the abstract and totalitarian plan which unveiled all the secrets of a town, showing its inner and outer workings.[29] Although views of cities continued to be made and diffused, increasingly larger numbers of plans were provided to satisfy scientific interest in the actual layout of foreign towns.

The best-known example of the consuming interest in maps that shows their connection to both war and cities is Tristram Shandy's uncle Toby, a wounded survivor of the battle of Namur. He begins his studies of siege and fortification in order to exorcise his fears and heal his wound, but soon they become a hobbyhorse. Not only does he become an expert on siege strategy, but his enthusiasm and devotion are so contagious that they contaminate the entire Shandy household and especially corporal Trim, his valet, who gamely offers to build for uncle Toby a model of the Namur siege in the garden.[30] Furthermore, his thinking about place as a location on a map becomes evident when, in answer to where he was wounded, he points to a bastion on Namur's circumvallation plan rather than the unfortunately strategic wound on his own body.

The reality of military architecture is illustrated in the treatises published toward the end of the seventeenth century by Nicolas de Fer (*Les Forces de l'Europe,* Paris, 1695), Daniel Bodenehr (*Force d'Europe,* Augsburg, c. 1708), and Galeazzo Gualdo Priorato (*Teatro del Belgio,* Frankfurt, 1673). These are a different kind of publication: although their purpose is still to educate in military architecture, their teaching is done in a historical manner, by providing examples of the great realized fortification projects rather than

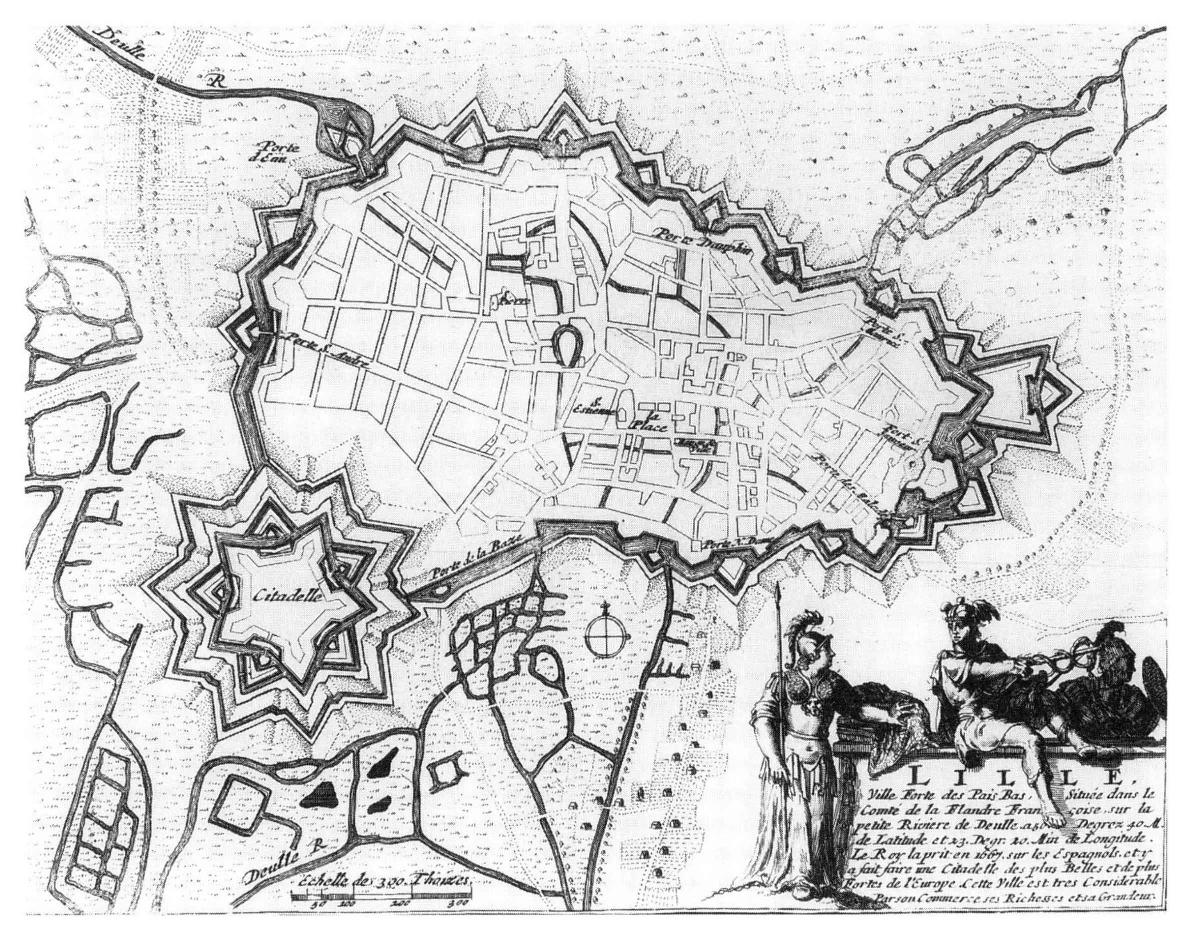

FIGURE 4.9. Nicolas de Fer, Plan of Lille. Copperplate engraving. From *Les forces de l'Europe, ou description des principales villes avec leur fortifications* (Paris, 1695–96). Photo courtesy of The Newberry Library, Chicago.

attempting to document a personal innovation and then persuade the community of military theorists of its originality. The illustrations that form the greater part of these works show the extent to which the ideas contained in the earlier treatises had been realized in bricks and stone. The plans and views of places as culturally and geographically distinct as Palmanova, Antwerp, Berlin, Turin, Lille, Coeworden, and Mannheim show that the widely accepted bastioned fortification surrounded most western European towns in the seventeenth century. They also document the proliferation of fortification outworks that isolated the city, turning it into an island, or placing it in its *assiette* according to Vauban. These illustrations show the city freestanding within a wide belt of land that has been cleared of buildings and trees. The open strip provided not only unobstructed sightlines for the defenders but also a field that could be mined against careless attackers during siege (figure 4.9). This menacing isolation of the city inside its mined belt and bastioned trace was effectively represented in the early histories of fortification, which thus publicized the terrifying character of seventeenth-century permanent fortification of cities.

These historical fortification treatises became invaluable for the education of the military architect and indispensable for the training of military strategists. Through their representation of an awesome military edifice they could act as a deterrent force. Their authors concluded the two-century-long effort that rendered military power scarily beautiful. Furthermore, because these books provided an aesthetic lens through which fortifications were to be viewed, they thus ennobled them with formal qualities that were eventually adopted also in civic architecture.[31] They promoted the work of military architects by codifying the disparate fragments of military defense, and they raised siege warfare to the level of science.

The earliest true plans (figure 4.10), whether manuscript or engraved, are inevitably fortification plans and studies (Milan 1497 ms., 1567 engraved; Lecce 1574 ms., Palermo 1575 ms., Florence 1520 ms., Turin 1558 ms.).[32] The relation between the bastioned wall, introduced at the end of the fifteenth century, and contemporary mapmaking has been pointed out by a number of architectural historians like Stanislas von Moos in his *Turm und Bollwerk: Beitrage zu einer politischen Iconographie der italienischen Renaissancearchitektur* (Zürich, 1974), historians of cartography like E. Pognon in his "Les plus anciens plans de villes gravées et les événements militaires" (*Imago Mundi* 22[1968]:13–19), and historians like John Hale in his "Warfare and Cartography" (typescript). Because the fortification had to be conceived as a whole, mapped out in its entirety, military needs prompted improvements

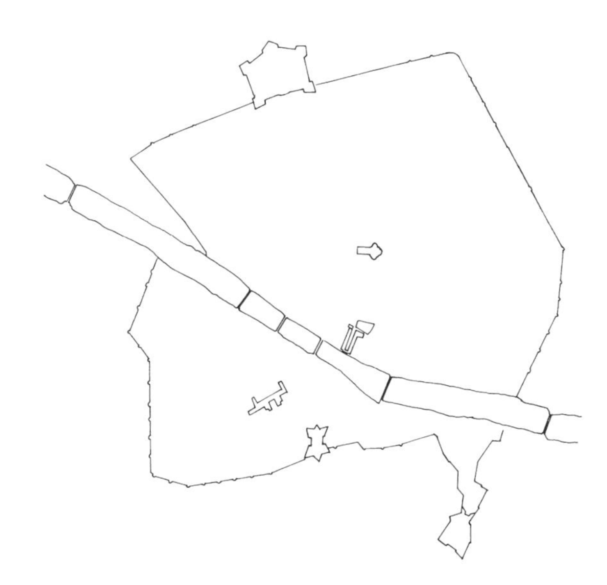

FIGURE 4.10. Tracing after a drawing by Baldassare Peruzzi, Fortification plan of Florence, c. 1520. Pen and ink. Prints and Drawings Collection, Uffizi, Florence.

in cartography. Thus the earliest plans coincide with the reconstruction of the cities' fortifications after the beginning of cannon warfare.

The seventeenth-century walls of Genoa, built as a triangle around the city along the ridge of the mountains, formed the definitive cartographic image of the city for more than two centuries, despite the fact that the fortifications could not be efficiently defended, nor could the vast area they enclosed be used for construction, since it was too steep (figure 4.11). There seems to be here an unexamined relation between the desire to express power and the role of military architecture in image making. In keeping with changing fashions, the image of Genoa as a city of palaces was to be replaced with an image of power achieved through military science. The construction of the extended walls coincides with the first accurate plans and views of the city, and both enterprises, cartographic and military,

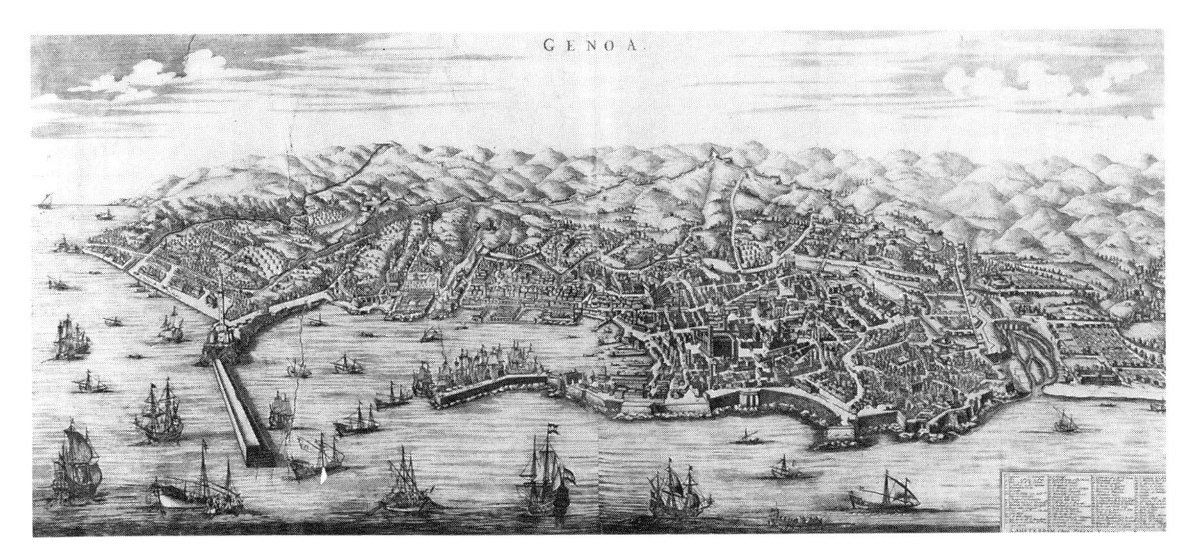

FIGURE 4.11. Pierre Mortier, Fortification plan of Genoa. Copperplate engraving, Photo courtesy of The Newberry Library, Chicago.

are evidently part of the pursuit of Genoa's nobility for a royal status for their city.[33]

Although quite often military studies show only the fortifications girdle, there are many plans, both manuscript and engraved, where the walls and the bastions are clearly the protagonists of the illustration, despite the streets, squares, and monumental buildings which are also delineated. The preferred point of view for Milan's cartography, as seen above, placed the citadel at the top center of the image. The military settlement thus dominated the civic life. The patent aggression of this image and the great number of military studies that show the city as a formidable war machine further establish the influence of cartography in the image of the city. The strategic value of the citadel is emphasized in its being rendered two to three times larger than it actually was in its relation to the city, significant in Montaigne's praise of Milan for its size and its awesome Castello.

Thus historic urban cartography is indelibly linked with military strategy and planning. If this postulate is accepted, then we also need to give central importance to the fortification and the defense of the city, which was the source of the prolific cartographic movement of the seventeenth century and came to influence the urban form and perception of the city.

NOTES

1. Pietro Cataneo, *I quattro primi libri* (Venice, 1554), 1: "la piu bella parte dell'Architettura certamente serà quella, che tratta delle città . . . le quali essendo modernamente offese dalle artiglierie, che no havevano gli antichi; non serà presontione la mia, se io mostrerò di edificarle altrimenti, per difenderle da quelle offese, alle quali essi non hanno potuto provedere, per non haverle havute al tempo loro." For a critical evaluation and reprint of the second edition of 1564, see *Trattati*, introduction by Elena Bassi and notes by Paola Marini (Milan, 1985); also Horst de la Croix, "The Literature of Fortification in Renaissance Italy," *Technology and Culture* 4 (1963): 30–50, and Hanno-Walter Kruft, "Vitruv, Festungsbau und Humanismus," in *Krieg und Frieden im Horizont des Renaissancehumanismus*, ed. Franz Josef Worstbrock, *Mittei-*

lungen der Commission für Humanismusforschung (Weinheim, 1986), 163–84.

2. For an introduction to the contributions to architectural and urbanistic theory made by Leon Battista Alberti (1404–72), Filarete, actually Antonio Averulino (c. 1410–c. 1480), and Francesco di Giorgio Martini (1439–1501), see Hanno-Walter Kruft, *A History of Architectural Theory from Vitruvius to the Present,* trans. Ronald Taylor et al. (London and New York, 1994).

3. Ludwig Heydenreich, "The Military Architect," in *The Unknown Leonardo,* ed. Ladislao Reti (London, 1974), 136–63.

4. This plan has received extensive scholarly attention; for recent questions of attribution, see Martin Kemp, *Leonardo da Vinci: The Marvellous Works of Nature and Man* (Cambridge, Mass., 1981), 228–30.

5. See Pierre Francastel, "Paris et la création urbaine en Europe au XVII siècle," in *L'Urbanisme de Paris et l'Europe 1600–1680,* ed. Piere Francastel (Paris, 1969), 9–37.

6. Simon Pepper, "The Meaning of the Renaissance Fortress," *Architectural Association Quarterly* 2 (1973): 21–28.

7. For the history of these sieges, see Christopher Duffy, *Siege Warfare: The Fortress in the Early Modern World, 1494–1660* (London, 1979).

8. Christine de Pisan, "Les Faicts d'armes et de chevalerie," ms. Bibliothèque Nationale, Paris, c. 1410; Valturio, *De re militari libri XII* (Verona, 1472); Niccolò Machiavelli, *Dell'arte della guerra* (Florence, 1521).

9. Francesco di Giorgio, "Trattato dell'architettura civile e militare" (dated variously between 1482 and 1495), ed. Corrado Maltese (Milan, 1967); Leonardo, "Codice Atlantico," (drawn between 1479 and 1513); for an examination of his drawings for fortresses, see Pietro C. Marani, *L'architettura fortificata di Leonardo da Vinci* (Florence, 1985).

10. For the Sangallo, see C. Huelsen, *Il libro di Giuliano da Sangallo* (Leipzig, 1910); Giancarlo Severini, *Architetture militari di Giuliano da Sangallo* (Pisa, 1970); *Antonio da Sangallo il Giovane; La vita e l'opera,* Atti del XXII Congresso di Storia dell'Architettura (Rome, 1986); and Nicholas Adams and Christoph Frommel, *Architectural Drawings of Antonio da Sangallo the Younger and His Circle* (Cambridge, Mass., 1994). For Michelangelo's contributions, see James Ackerman, *The Architecture of Michelangelo* (London, 1961), chap. 5; and Vincent Scully, "Michelangelo's Fortification Drawings: a Study in the Reflex Diagonal," *Perspecta* 1 (1952): 38–45.

11. For the history of the polygonal bastion's development, see John R. Hale, "The Early Development of the Bastion: An Italian Chronology, c. 1450–c. 1534," in *Europe in the Middle Ages,* ed. Hale et al. (Evanston, Ill., 1965), 466–94.

12. See John R. Hale, *Renaissance Fortification: Art or Engineering?* (London, 1977).

13. Montaigne, *Oeuvres Complètes* (Paris, 1962), 1334–35; *Il conte Fulvio Testi alla corte di Torino negli anni 1628 e 1635,* ed. D. Perrero (Milan, 1865), 126; Joan DeJean, *Literary Fortifications* (Princeton, N.J., 1984), 21–23.

14. The classic monograph on Dürer is Erwin Panofsky, *Albrecht Dürer* (Princeton, N.J., 1955), but his fortifications treatise is analyzed in William Waetzoldt, *Dürers Befestigungslehre* (Nuremberg, 1916), and *Dürer and His Times* (London, 1950), 220–26; see also Martin Biddle's introduction to *Etliche Underricht zu Befestigung der Stett, Schloss, und Flecken* (1527; rpt. New York, 1972). For Vauban, see Reginald Blomfield, *Sébastien le Prestre de Vauban, 1633–1707* (London, 1971); Pierre Lazard, *Vauban, 1633–1707* (Paris, 1934); Michel Parent and Jacques Verroust, *Vauban* (Paris, 1971); and F. J. Hebbert and G. A. Rothrock, *Soldier of France: Sébastien le Prestre de Vauban, 1633–1707* (New York, 1989).

15. For complete bibliographic references on these authors of military treatises, see my *Military Architecture, Cartography, and the Representation of the Early Modern European City* (Chicago, 1991).

16. "The Argument of Some Military Title Pages of the Renaissance," *Newberry Library Bulletin* 6, 4 (1964): 91–102.

17. Quoted from Arthur Koestler, in Paolo Rossi, *La scienza e la filosofia dei moderni: Aspetti della rivoluzione scientifica* (Turin, 1989), 90.

18. Rossi, *La scienza,* 91.

19. For Lorini, see Carlo Promis, "Biografie di ingegneri italiani dal secolo XIV alla metà del XVIII," *Miscellanea di Storia Italiana* 14 (1874): 650–75; for Errard, see David Buisseret, "Les ingeniéurs du roi au temps de Henri IV," *Bulletin du Comité des Travaux Historiques et Scientifiques: Section de Geographie* 77 (1964): 13–84.

20. For Freitag, see Alessandro Biral and Paolo Morachiello, "Filosofo, soldato, politecnico," in Alessandro Biral and Paolo Morachiello, *Immagini dell'ingegnere tra quattro e settecento* (Milan, 1985), 62–65.

21. For Marolois, see ibid., 51–57; for a discussion of Vauban's *Cahiers de charge,* see Hebbert and Rothrock, *Soldier of France,* 72–94.

22. For discussion and further bibliographic references on Coehoorn and Dögen, see *Architekt und Ingenieur, Baumeister in Krieg und Frieden,* ed. Ulrich Schütte (Wolfenbüttel, 1984), 376–77, 368–69.

23. See Leone Ebreo *Dialoghi d'amore* (Rome, 1535) and Pico della Mirandola, *Opere* (Bologna, 1496).

24. Among the particular studies of these citadels, see

Hugo Soly, "De Bouw van de Antwerpse Citadel (1567–1571): Sociaal-Economische Aspecten," *Belgisch Tijdschrift voor Militaire Geschiedenis* 21, no. 6 (1976): 549–78 (with French resume); Charles van der Heuvel, "Bartolomeo Campi, Successor to Francesco Paciotto in the Netherlands: A Different Method of Designing Citadels," in *Architetti e ingegneri militari italiani all'estero dal XV al XVIII secolo,* ed. Marino Viganò (Rome, 1994), 153–67; Giovanni Fanelli, *Firenze, architettura e città* (Florence, 1973), 262–63; and Bruno Adorni, *L'architettura farnesiana a Parma, 1545–1630* (Parma, 1974), 147–63.

25. For a study of the role of the citadel, see Paolo Marconi, "Una chiave per l'interpretazione dell'urbanistica rinascimentale: la cittadella come microcosmo," in *Quaderni dell'Istituto di Storia dell'architettura* (Rome, 1968); for the ambivalent relations between the population and the government that it provoked, see Niccolò Machiavelli's chapter on the citadel in his *Arte della guerra* (see n. 8).

26. Galasso Alghisi, *Delle fortificationi libri tre* (Venice, 1570).

27. In a memorandum, composed with the aid of Castiglione, to Pope Leo X, first published by Vincenzo Golzio, *Raffaello nei documenti, nelle testimonianze dei contemporanei e nella letteratura del suo secolo* (Vatican City, 1936); Barbaro's suggestions for architectural representation are sensitively discussed by David Rosand, *Painting in Cinquecento Venice* (New Haven, Conn., 1982), 178.

28. Galileo's *Breve istruzione all'architettura militare* (1593) and *Trattato di fortificazione* are analyzed by Catherine Wilkinson Zerner in "Renaissance Treatises on Military Architecture and the Science of Mechanics," in *Les traités d'architecture de la Renaissance,* ed. Jean Guillaume (Paris, 1988), 467–76.

29. See Kim Veltmann, "Military Surveying and Topography, the Practical Dimension of Renaissance Perspective," *Revista da Universidade de Coimbra* 27 (1979): 263–79.

30. Laurence Sterne, *The Life and Opinions of Tristram Shandy, Gentleman* (1760; Oxford, 1983), 2: 65–122.

31. For the influence of military architecture on civil architecture, see Christian Otto, "L'Architecte et l'ingenieur: Neumann et la Residenz de Werneck," *VRBI* 11 (1989): 71–84, and Wilhelm Waetzoldt, *Dürer and His Times* (London, 1950), 224: "Dürer's study of the theory of military architecture is a presage of a new development which was to appear in the period of the German Baroque—military engineering as a basis of monumental architecture, with military engineers and artillery officers (Speckle, Dientzenhofer, Neumann) as masters of sacred and profane architecture."

32. For a discussion of the earliest manuscript maps of Italian towns, see, by specific city, the series *Le città nella storia d'Italia* (Bari), edited by Cesare de Seta, and my review of the volumes published prior to 1988 in "La storia delle città: testi, piante, palinsesto," *Quaderni Storici* 67 (1988): 223–56.

33. For the cartographic and symbolic transformation of the city, see the study by Ennio Poleggi and Paolo Cevini, *Genoa* (Bari, 1981).

FIVE

Modeling Cities in Early Modern Europe

David Buisseret

Introduction

The collection of city models now housed in the Musée des Invalides in Paris is probably the most famous, and certainly the most extensive, in Europe. It was begun in 1668, and by the middle of the nineteenth century comprised about 125 large models, showing many towns and forts both in France and outside her borders.[1] Various studies of this *Musée des Plans-Reliefs* have been published during the past thirty years, and we shall come back to it in the second part of this chapter. However, what has not been well understood until very recently is that this French collection was only one among many in early modern Europe, and by no means the earliest.[2]

The History of City Models in Europe

The medieval cathedral builders sometimes made models of the vast structures that they proposed to build.[3] It seems to have been somewhat in line with this tradition that the military architect Basilio dalla Scuola in 1501 displayed in Venice a wooden model of a fortress, with the aim of showing "what is being done in France, Italy, . . . Germany and elsewhere."[4] The same architect in 1521 made a model of Rhodes, which was in imminent danger of siege by the Turks; this work was sent to Pope Leo X.[5]

Meanwhile in Germany the emperor Maximilian I was also having such objects made; in 1514, for instance, Adolf Daucher made a model for him of the Luginsland Tower in the fortifications of Augsburg.[6] Models and maps are different ways of expressing topographical reality, and it can be no accident that models of this kind emerged from circles in which there was also precocious use of maps. The Venetians were among the first to use maps in controlling their *Terraferma,* and Emperor Maximilian was renowned among his contemporaries for his cartographic skill; he could, it was said, make maps out of his head of any part of his extensive domains.[7]

The papacy had long been interested in cartography, and, in 1529, when Pope Clement VII had decided to besiege Florence, he gave orders to his agents there for a detailed model of the city to be made. The work took several months and was executed in cork for lightness. "The final model was broken down and packed in boxes which were surreptitiously carried out of Florence, hidden in bails of wool destined for Perugia. . . . The Pope constantly used this model during the siege, keeping it in his room and using it to check the information coming in about the site of the camp, the location of fighting and so forth; in a word, everything to do with the siege. This unusual and wonderful work proved highly satisfactory."[8]

Other Italian rulers also became aware of the potential of models. Giorgio Vasari painted Cosimo de' Medici, grand duke of Tuscany, during the 1580s, studying a plan and model of Siena; in their caption to this painting, Pepper and Adams observe that "there are other Sienese images which lead one to believe that models were not uncommon."[9] In Venice, a regular program of model building seems to have begun in 1571, designed to show the Venetian forts in the Levant.[10] These models included such sites as Candia, Spalato, and Famagusta, and were made of wood. They eventually numbered twenty, of which eighteen still remain in the Museo Storico Navale in Venice. Later, other Italian centers such as Bologna and Naples also produced models.[11]

In Germany, the initiative begun by Emperor Maximilian was followed at various sites in Bavaria.[12] In 1540, for instance, the painter Hans Behaim made a model of Nuremberg, which Veit Stoss had also modeled in 1530. Between 1560 and 1563 Hans Rögel produced one of Augsburg; Behaim and Rögel seem to have been self-taught, as was Jakob Sandtner, who in 1568 presented a model of his hometown, Straubing, to Duke Albrecht V. The duke was interested in representations of his lands—it was he who commissioned the magnificent map by Philip Apian—and he ordered four more models from Sandtner. They covered the most important Bavarian administrative centers of Munich, Landshut (1571), Ingolstadt (1572), and Burghausen (1574); Sandtner also modeled Rhodes and Jerusalem, the latter before its capture by the Moslems. Five of his models are still preserved in the Bavarian National Museum in Munich.[13]

By the middle of the sixteenth century, it may indeed have been quite habitual for models to be made when extensive works were envisaged.[14] In 1550, for instance, when the emperor Charles V was contemplating the construction of a citadel at Siena, he commissioned Giovambatista Romano to make a model of the city and its fortifications, which with other information was taken to him for a decision; he then confirmed one of the suggested sites.[15] His successor, Philip II, was acutely aware of the need to make maps and plans of all his possessions, and so it is not surprising to learn that he also commissioned a set of wooden models.[16] His engineers probably made them to set out potential works, as in 1597 when Cristóval de Rojas made a model of the proposed works at Cartagena.[17]

We have few references to model making from seventeenth-century Spain, but in 1723, no doubt as part of the Bourbon imitation of things French, Philip V commissioned some models, and in 1777 Charles III formed a *Cabinet des Plans-Reliefs*, under the direction of the Fortifications Service. It had great ambitions, but in the end completed only a model of Cádiz, now in the Museo Histórico Municipal there.[18]

The English do not seem to have been active in model making during the sixteenth century, though they were certainly aware of developments on the continent. When in 1570 Sir Humfrey Gilbert proposed the foundation of an academy in London for the education of young nobles, his syllabus included not only the ancient languages, divinity, law, medicine, and natural science but also military studies, including instruction on how to "draw in paper, make in modell and stake out all kinds of fortifications."[19]

This "making in modell" was particularly important in the case of fortifications, where each part needed to support the others, as part of a general fire control system. Writing forty years or so after Gilbert, Simon Stevin in chapter 2 of his *Fortification* suggests that for each work, three models ought to be made.[20] The first would show the whole system, including the streets, houses, and churches that it was protecting. The second would show a half-bastion in detail, and the third would set out part of the wall and two bastions, to ensure that the distances and angles were correct. It seems likely that many engineers of the seventeenth and eighteenth centuries followed Stevin's advice, though most of their models have been lost.

One extensive collection that has survived (in the Royal Military Academy at Breda) is the one covering fortresses in the United Provinces and in Dutch possessions overseas, for which both the *stadhouder* and the estates of Holland commissioned detailed models.[21] These seem to have been made of wood and were relatively small. In the early seventeenth century they were commissioned by Maurice and Frederick Henry, and others were made in the time of William III. They eventually numbered about forty-eight, as we learn from the inventory made when the French carried them off in 1810; they came back to the Netherlands after 1815. These Dutch models seem to have been severely practical, limiting their coverage to the actual fortifications; they probably had very little celebratory function.

The chronological incidence of modeling was different to the south, in the Spanish (afterward Austrian) Netherlands.[22] Here we have no information for the seventeenth century, but between 1740 and 1750 Charles-Alexandre de Lorraine assembled a large and coherent collection of seventy-one models, covering not only the Spanish Netherlands (twenty-nine models), but also Hungary and the Balkans (twenty-four), northern Italy (thirteen), and the Rhine frontier (five). They were kept in the Palais d'Orange, at Bruxelles, and showed the strongpoints in the theaters in dispute with the Turks in the East and the French in the West.

These models were relatively small, measuring about a meter square, and may not have shown much internal detail; only Vienna and Luxembourg rated somewhat larger examples. We do not know if they were shown to distinguished visitors, or if they were strictly a working collection. Charles-Alexandre died in 1780, and his collection passed to his nephew, Emperor Joseph II. The latter had visited the Paris collection of *Plans-Reliefs* in 1777 but does not seem to have been much impressed by their utility; at all events, he ordered that his uncle's collections should be burned, and this order was carried out. It seems extraordinary that Joseph II could thus destroy what seems to us a historic source—including two models of his own capital, Vienna—but the emperor was a headstrong and authoritarian character who despised his uncle and wanted to break with the past.

As we shall see, the French maintained their collection throughout this period and even added to it. Among their models at this time were some overseas cities, including Quebec. But in 1808 another plan of Quebec was prepared in Canada itself, a rare example for the time of a model being conceived and built outside Europe. Its authors were Jean-Baptiste Duberger, surveyor and draftsman, and John By, engineer,[23] they worked in anticipation of the conflict with the United States that actually broke out in 1813. The model was initially more than ten meters long and was built to the same kind of detail as the French models, showing all the topographical details, buildings, fortifications, and so forth. In 1810 it was taken to England, where decisions about the fortifying of Quebec were then made; eventually it lost its functional value and in 1908 was returned to Canada, where it is now most elegantly displayed by Parks Canada at Artillery Park in Quebec.

This model constitutes a fine example of the way in which such works can bring the past to life for a wide public. A light show lasting twenty-four minutes reproduces a normal day of twenty-four hours in old Quebec, using lighting to pick out and identify important buildings and areas of the city. Exhibits of this kind are wonderfully effective in giving visitors (and residents) a strong sense of the outline of a historic site, and are all the more evocative if the model, as in the case of Quebec, was made almost two hundred years ago.

Models of cities and fortresses continued to be made during the nineteenth century, when, for example, an extensive collection was assembled by the Prussian engineer Alexander von Zastrow (c. 1800–1875).[24] The advent of contour maps diminished to some extent the importance of models, though they continued to be made for ventures like the digging of the Suez Canal.[25] Models were not much used during the First World War, whose trench landscape could easily be delineated on flat maps. With the coming of the Second World War, though, models again came into their own. From 1939 onward, the Swiss developed an active *Service des reliefs de l'Armée*,[26] and both U.S. and British forces developed huge model-making capacities, organized respectively in the Model Making Detachment of the Corps of Engineers and in V-Section of the Allied Central Interpretation Unit, Royal Air Force.[27] They made over three hundred models, of which the most remarkable was probably the one of the D-Day landing beaches. Measuring over sixty feet long, it covered the coast from Caen westward to Cherbourg at a scale of 1:5,000, and was intensively used during the final planning and briefing.

After the war, these skills found applications with commercial publishers, so that firms like Nystrom in Chicago brought out whole series of relief models, often at the level of individual states. The advent of digitized imaging has probably reduced the importance of topographical modeling, but as late as 1987 a working group of the International Society for Photogrammetry and Remote Sensing held a conference in Denmark titled "Progress in Terrain Modeling."[28] These more recent models tend to show not so much cities as countrysides, but schools of architecture continue to train their students by making huge models of towns, which are also commissioned by some tourism centers to give visitors a quick but accurate feel for their urban areas. Some cities, like Washington, D.C., have also developed tactile models of their buildings and streets for the use of the blind.[29] The making of models, which seems to have begun in the early sixteenth century, has thus had a long history, one that is not yet over.

The French Example: The History of the *Plans-Reliefs*

According to Girolamo Maggi, an Italian engineer in the service of the kings of France, he had during the 1550s made many drawings and models of fortresses for Henri II (1547–59).[30] However, none of these works survives, and we curiously have no mention of modeling in the time either of Henri IV (1589–1610) or of Richelieu (1624–42), both of whom were prominent in encouraging other forms of mapping.

Serious modeling in France seems to have begun with the military engineer Allain Manesson-Mallet (1630–1706), author of *Les Travaux de Mars ou l'art de la guerre* (Paris, 1685). He was then a mathematician in the service of Louis XIV but had earlier served the king of Portugal, for whom he had built numerous fortifications.[31] As he was returning to Portugal in 1663, he had been asked to make a model of Pignerol for presentation to the king. Manesson-Mallet puts it like this: "I confess that I got the idea from a model made by an Italian engineer, but in so doing I inspired many others in France, who have since carried the art forward."[32] As so often seems to be the case, this was an idea that was rather in the wind at this time, for two years later the king received from François Andréossy, an engineer working on the Canal du Midi, a model of the city of Narbonne.[33] The king seems to have been much taken by these models and probably resolved at this time to commission more of them.

An occasion for this soon came up, for in 1668, by the treaty of Aix-la-Chapelle, France acquired the frontier towns of Aire-sur-Lys, Ath, Audenarde, Bergues, Charleroi, Courtrai, Douai, Lille, Menin, and Tournai. These cities needed to be fortified, and the best way of showing to the king what needed to be done was to make a model. Other occasions arose for models to be constructed; from 1685 onward, for instance, French policy increasingly turned toward maritime affairs, with the result that plans of seaports began to be numerous.[34] By 1698 there were 144 models, all of which had been scrutinized by the king, who during the 1690s generally set aside Monday afternoons for work of this kind.[35]

In these early years, each model was made on the spot and then transported (often with much difficulty, for they were bulky and fragile) to Paris, where, after the king had examined it, the model was stored in the palace of the Tuileries.[36] This procedure was followed until 1706, when the space in the Tuileries must have been getting rather crowded; at all events, from that year onward until 1715 the models were steadily moved across to the *Grande Galerie* of the Louvre, where they could be displayed most effectively in that huge long room built by Henri IV in the early seventeenth century. An exceedingly rare scene shows Louis XV with some members of his staff examining both a map and, no doubt, the corresponding model.[37]

From the time that the collection was transferred from the Tuileries, it acquired the additional function of being a prestige object, a sort of massive *Wunderkammer* to be shown to amazed foreign visitors (though not, if possible, to their leading engineers, who might learn altogether too much about some sites). Peter the Great visited the collection in 1717 and is said to have formed a similar one in Russia, though little is known about it.[38] In 1721 the Turkish ambassador, Mehmed Efendi, was shown 125 models and was immensely impressed by their military potential as well as by the scale of the work.[39] The duke of Parma viewed the models in 1763 and then commissioned several models for his own purposes.[40] The last visitor of note was Joseph II, who toured the *Grande Galerie* in 1777, but who, as we have seen, was not impressed.[41]

The models were still numerous at that time, in spite of the great problems of maintenance (about which more later), but they were beginning to lose their appeal. Vauban had always thought that they were a waste of time for engineers who could read maps, and the introduction of contour lines about 1760 meant that relief could (in theory) be conveyed by this accurate and much cheaper means. During 1776 and 1777, then, the models were transferred to the attics of the Invalides, which had been used to store grain. The work was immense, for at least one thousand cart trips were needed. It was also delicate, and perhaps not carried out with sufficient care, for twelve models were destroyed in the course of this move.[42]

The models were much less grandly housed in their new home and might have been simply neglected. But during the revolutionary years, in 1791, a special *Comité des Fortifications* was formed, responsible among other things for the engineering archives. The director from 1793 onward was Colonel Carnot (brother of the famous Lazare Carnot), and he had an acute feel for the value of the models, which he said could "offer information about all questions relative to terrain."[43] Moreover, Napoleon himself was persuaded of their value, observing that "there is no better map than these relief models."[44] So the collection was not only preserved but even extended, with models of places like Toulon and Brest added and others brought in from conquered rulers like the king of Sicily and the Dutch *stadhouder.*

After Waterloo, the collection lost the models that had been seized by Napoleon. It also lost nineteen of its own models to the victorious Prussians; they carried them back in triumph to the *Zeughaus* (armory) in Berlin, where all except one were destroyed during the Second World War. However, if this was lamentable for the collection, it is a testimony to the continuing value of the models for military and prestige purposes. Even between 1830 and 1860, models continued to be added, with Antwerp, Rome, and Constantine joining the earlier set. The models were now largely used for display, being exhibited, for instance, during the 1867 exposition.[45] As late as 1870 the models of Metz and Strasbourg were brought up to date, but the mention of Metz—dread word for the late-nineteenth-century French—reminds us that the war of 1870, which seemed to show the uselessness of *places fortes,* dealt a final blow to the collection of models as a military tool.[46]

Henceforward they were simply an encumbrance in the attics of the Invalides, and in 1885 they were transferred to the *Service Géographique de l'Armée.*[47] A period of peril then ensued when they were little prized, and some indeed were destroyed merely to make space. However, in 1927 the survivors were classed as *monuments historiques,* and so their continued existence became more likely. During the Second World War they were evacuated, either to Sully-sur-Loire or to Chambord, or perhaps to both.[48] After the war they were then returned to the Invalides, where they have come to be increasingly appreciated as a remarkable source for understanding the cities of the *ancien régime.*[49]

Figure 5.1 maps the models surviving in 1990. There had been some notable losses: not only the models destroyed in the bombing of the *Zeughaus,* but also three Canadian models, of Montreal, Quebec, and Louisbourg, which perished during the transfer of 1776–77. Still, the remaining models form an impressive collection, covering not only the towns on France's borders but also cities in the Low Countries and in Italy. Their scale varies, but most are at a scale of about 1:600, which in the *ancien régime* was one *pied* to one hundred *toises,* or in English terms about a foot to two hundred yards.

The advantage of models over maps, then as now, is that they are immediately accessible, and allow the onlooker to take in any possible angle. As Roux puts it, their role was "to make visible what nobody then could see."[50] In fact, images taken from them are often extraordinarily similar to modern aerial photographs of the same site; by the magic of their art, the engineers of Louis XIV in effect provided their master with an aerial view before that was technically possible. Of course, the models also had grave disadvantages, the chief of which were that they were expensive to make, difficult to transport, and required both careful storage and constant maintenance. Still, when a large number of cartographically unsophisticated people need to be "shown" a city or a landscape, they are still the tool of choice, as is shown by their use in the Second World War. It is possible that the coming of virtual reality in electronic images has finally removed their raison d'être, but this is by no means certain. Models can now be made in plastic very

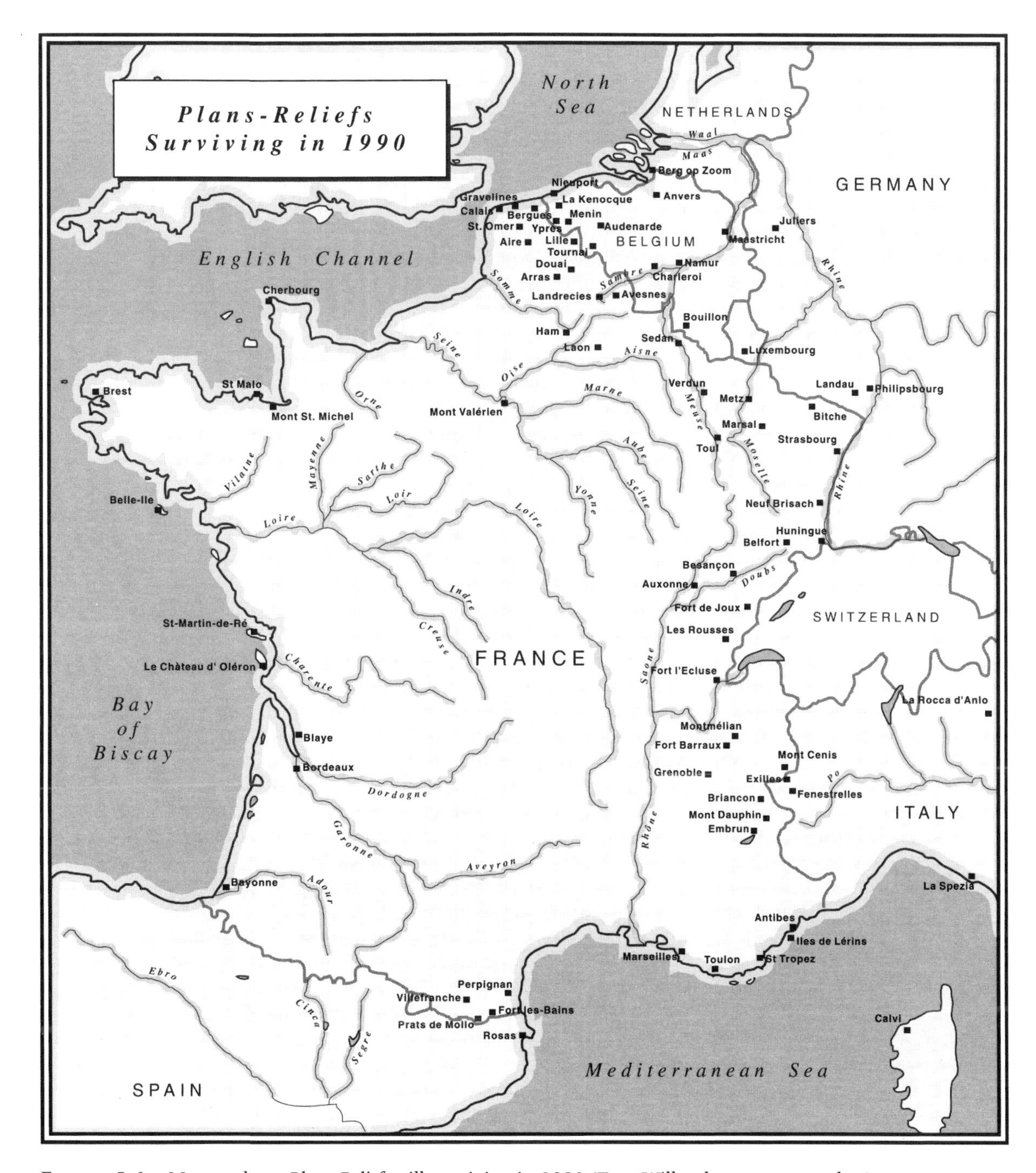

FIGURE 5.1. Map to show *Plans-Reliefs* still surviving in 1990 (Tom Willcockson, cartographer)

FIGURE 5.2. Model showing the town and citadel of Montmélian after the French siege in 1691

cheaply, and they have become the inexpensive alternative to newer, much more costly forms of electronic imaging.

We have noticed in passing that the function of the Paris models varied over time. At first they were strictly working tools, made in order to let the king inspect the state of the fortifications, and perhaps in some sense to persuade him to undertake costly works. Later, they extended their coverage, so as to enable the king to follow and even direct sieges of enemy towns. After their transfer to the Louvre, they still served as a tool for staff officers and for instructing gunners and engineers, but they were now also part of the policy of prestige, a wonderful way of impressing potential allies (and enemies) with the grandeur of France, expressed in this collection showing so many of its towns. They were impressive not only in their rich variety but also in some cases for a more direct lesson; the ruins of

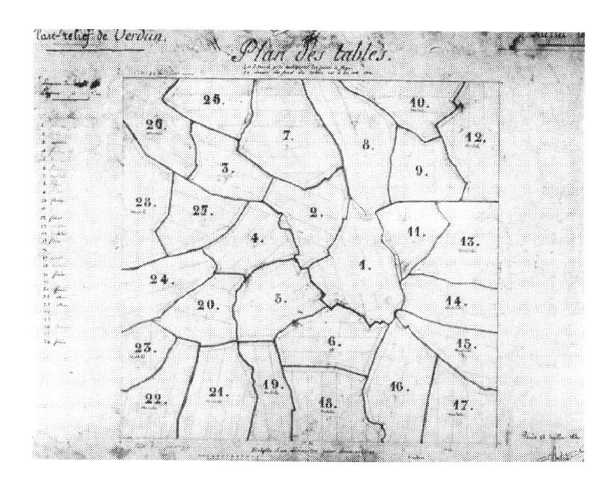

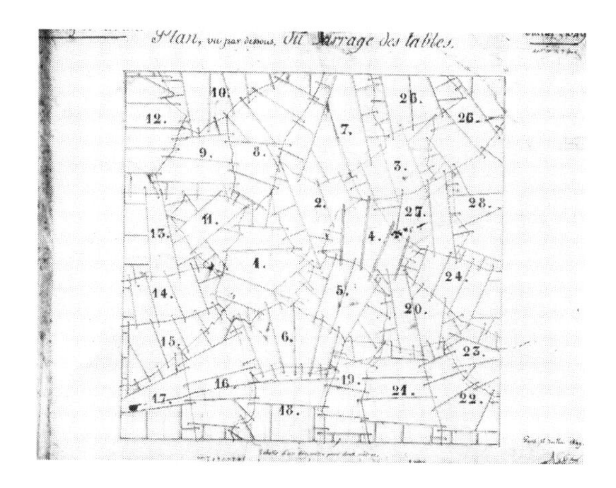

FIGURE 5.3. The assembly plan for the tables and the supporting frames of the Verdun model (1848–58)

Montmélian, graphically modeled, would certainly encourage second thoughts on the part of any foreigner likely to challenge the French artillery. Figure 5.2 shows what Montmélian looked like after General Catinat had finished bombarding it in 1691; one bastion has been ruined, and the whole town is destroyed.

Finally, the models eventually came as well both to have a strong aesthetic appeal and to offer a remarkable historical source. The aesthetic appeal must have been most powerful when they were exhibited in the *Grande Galerie,* but they are impressive as art objects even in their cramped quarters under the roofs of the Invalides. Their potential as a historical source becomes obvious to anybody who spends a few hours flying, in effect, over the towns and countryside of early modern Europe that they represent.

Although the collection seems to be homogeneous, in fact its models date from widely differing periods, and were constructed according to principles which varied over time. Over the two centuries from 1660 to 1870, different techniques of construction were used, though the basic elements remained the same. The models were made up of a number of "tables," attached by special rods and mounted on special stands. Sometimes these assemblies were almost unimaginably complex; the plan for Verdun, for instance, made in 1848–58, had twenty-eight separate tables, as Figure 5.3 shows. The view from underneath is, if anything, even more daunting, as it shows the complex assembly of rods needed to hold the work together. Until about the middle of the eighteenth century, the tables were constructed on the spot and then taken back to Paris; after that time they were constructed in special central workshops.[51]

From the start, the models attempted to reproduce reality as faithfully as possible; at the end of the seventeenth century, Vauban, for instance, insisted that he wanted each individual house to be correctly drawn, and as time went by the models became on the whole more and more accurate. Man-made objects were reproduced as closely as possible, and botanical features were shown as they would appear at the time of maximum flowering, in May or June. All kinds of material were used in the construction, including wood, silk, sand, and paper; these were then painted in realistic colors.

The foundation tables were very solid, but the modeling materials were equally fragile. Some fell down if the tables were banged; others rusted and collapsed; others again lost their color or went black under the influence of humidity and sunshine. All were subject to the fall of dust, which eventually covered them with a thin layer hard to remove. The models required total restoration every fifty years, and this was a highly skilled task, involving the overhaul of the underlying tables, the "planting" of fresh vegetation, the reconstruction of houses, and so forth. Needless to say, many models were sadly neglected at times when the government was uninterested in their fate.[52] The result of all this labor was to produce a set of images that in essence still allow us to visualize large areas of early modern France with a detail that would be impossible to generate from any other sources. We see the towns not only with their street plans, as if on a conventional map, but also as they are located in the surrounding countryside, with its farms, tracks, and often specialized agriculture; we also see the adjacent roads, some of which date back to the Romans, and others of which were the work of the eighteenth-century engineers.

Sometimes the towns astonish us by their apparent spaciousness and their abundance of green spaces; more often we are struck by the extremely crowded houses, especially in the great cities of the north like Tournai and Lille.[53] Figure 5.4 is a detail from the model of Tournai, showing the five-towered cathedral of Notre Dame and the thickly clustered adjacent houses. Looking at these models, it is easy to imagine how in these crowded quarters disease could spread very rapidly; it is also easy to appreciate the way in which nineteenth-century urbanists like Baron Haussmann felt that they had ruthlessly to attack these centers of disease and, sometimes, of social unrest.

FIGURE 5.4. Detail from the model of Tournai (1701)

In every case, the small overall size even of major centers like Strasbourg and Ypres is striking. On the model made under Louis Napoleon between 1856 and 1863 (Figure 5.5), we see Strasbourg still confined behind the fortifications constructed by Vauban. The cathedral and the Place de Broglie (to the left) stand out well, as does the newly channeled Rhine. A new railroad line comes in from the left, picking its way between the outworks to end in a large station, across the river from the main town. The fortifications, extending far into the countryside, comprehend not only the historic center on its island

FIGURE 5.5. The model of Strasbourg (1836)

but also three great suburbs, in some of which we see barracks for the soldiers who manned the defensive works.

As an example of a smaller town we might take Bayonne, at the confluence of the Adour and the Nive, in southwestern France (Figure 5.6). This border town, whose outline has changed very little, consists of three elements: the four-bastioned Citadelle du Saint-Esprit on the left, the old town on the right, and the "new town" built by Charles VII across the river Nive from it. The fortifications are particularly prominent, for most kings from François I onward worked at strengthening them, and Bayonne remained until very recently an important stronghold.

The models show us not only general images of towns but many examples of individual buildings. The model of Perpignan, for instance, made in 1686 (Figure 5.7) shows the church of Saint-Jacques as it was incorporated into the bastion built by Charles V., and the remarkably complicated system of gates and approach roads.

Auxonne was once on the frontier with Spanish Burgundy and so was heavily fortified. In Figure 5.8 we see part of the citadel constructed by Louis XI, with a small fortified work of the seventeenth century in front of it. Note the embra-

FIGURE 5.6. The model of Bayonne (1819)

FIGURE 5.7. Detail from the model of Perpignan (1686)

FIGURE 5.8. Detail from the model of Auxonne (1677)

sures for cannon on both the medieval towers and the new work. This model was made in 1677 and restored in 1771 and 1906; it no doubt shows Auxonne much as Napoleon Bonaparte would have known it when he was a young artillery lieutenant there in 1787–91. Saint-Omer passed into French hands by the Treaty of Nymwegen (1678), and the model shown in Figure 5.9 was made in 1758. In the foreground is a bastion dating from Spanish times and in the background two churches. Between them is a circular mound which was the site of a medieval castle, though it has recently been shaped to receive cannon. The models often show features like this, which mark a transient phase in the development of the fortifications. Most such sites, like this one, have since been lost in the development of French towns.

The model of Audenarde, made in 1747 (Figure 5.10), shows the thirteenth-century *Château de Pamele,* which was destroyed in the middle of the eighteenth century, shortly after the model was made. It lay alongside an arm of the River Escaut and so controlled traffic on this river from the town, famous for its cloth production. The boat looks rather crudely made, but it too probably gives a good impression of the type of craft

FIGURE 5.9. Detail from the model of Saint-Omer (1758)

FIGURE 5.10. Detail from the model of Audenarde (1747)

used on the Escaut at this time; this model is in fact remarkable for its elegance and precision.

Our final city detail comes from Besançon, captured by Vauban in 1647 and substantially refortified (Figure 5.11). It long continued to be a major French fortress on the Swiss frontier and has changed relatively little. This detail from the model made in 1722 shows Fort Griffon, a self-contained fortress with barracks, chapel, magazine, and officers' quarters. The model is remarkable for its skillful use of silk in delineating trees. Today the old fort has become a school for girls.

Specialized crops are easily identified in the countryside around many of the towns. Outside Toulon, in a model made at the end of the eighteenth century, we see a few wheat fields and a good deal of arid pastureland. The roads are lined with trees, perhaps a reflection of the attempt to grow elms for artillery wood alongside the main roads in France. The farms outside Audenarde have thatched roofs, and a great variety of crops is growing in the small fields; perhaps we can recognize wheat and some apple orchards. The plan of Brest (Figure 5.12) was made quite late, in 1811, and is remarkably precise. Here we see a detail from it showing the *bocage*, a countryside divided into many small fields separated by hedges. Looking at this marvelously expressive model, we cannot help wondering if the Allied commanders in 1944, who had such excellent models of the beaches, had any such models of the Norman *bocage*, through which their armies made their way with so much loss and difficulty.

Models like the one of Philipsbourg show much more open country and often delineate a

FIGURE 5.11. Detail from the model of Besançon (1722)

FIGURE 5.12. Detail from the plan of Brest (1811)

pattern of roads shaded by trees, probably the omnipresent poplars in the north of France. Outside Toul there are vineyards now long disappeared, and outside Verdun there are hop fields. Many cities, like Toulon and Perpignan, have truck farms near them (Figure 5.13), and some, like Douai and Tournai, contain them within their walls. Around the cities of the north are often polders, and around those of the west coast salt pans; in short, the models give us a remarkable overview of rural economic activity.

They also show us some industrial processes. The Strasbourg plan, updated between 1856 and 1863, shows the new railroad system, including its roundhouses (Figure 5.14). These were the maintenance hangars of the whole system, into which locomotives were retired each night in order to be cleaned and fueled for the next day. On the model of Metz, built between 1821 and 1825, we see the very distinctive shape of the tanneries, alongside one of the arms of the Moselle (Figure 5.15). These distinctive buildings have now completely disappeared, but it is easy to imagine the degree of pollution that could be

FIGURE 5.13. Detail from the model of Toulon (1796–1800)

FIGURE 5.14. Detail from the model of Strasbourg (updated 1856)

caused in a relatively small river by such a concentration of industrial activity.

In one or two instances, two models exist of the same town at different dates. This is the case for Strasbourg, for instance, where the model of 1727 was confiscated by the Prussians in 1815, and another one made in 1863. The later model is now at the Invalides, while the earlier one was returned by the Germans in 1903 and is now at the historical museum in Strasbourg. In this particular case the contrast is striking, for we have here a view of the town long before the railway age and then at the very dawn of it.

Of course, the models are also of extraordinary interest to historians of fortification. The model of Landrecies, made in 1723 (Figure 5.16), shows the medieval town center, protected by the wall and small bastions constructed by the engineers of Charles V in the first half of the sixteenth century. After 1668 Vauban surrounded this original fortress with much larger works, including one which stretches far out into the countryside. Landrecies was a very strong site, made stronger by its extensive flood defenses; curiously, nothing seems to remain of these structures today.

The model of Nieuport, built in 1698 (Figure 5.17), tells a different story; here there is a medieval planned town, protected by a wall raised by the fifteenth-century Burgundian dukes. Onto this wall were grafted at various times a set of bastions, though the whole system looks rather inadequate, and the town in fact depended right down to the First World War on flooding the sur-

FIGURE 5.15. Detail from the model of Metz (1821–25)

FIGURE 5.16. Vertical view of the model of Landrecies (1723)

rounding countryside for its protection. Nieuport was destroyed between 1914 and 1918, and this model remains the most extensive evidence of what it once looked like.

Whereas the fortifications of Landrecies and Nieuport have almost completely disappeared, at towns like Neuf-Brisach the fortifications survive much as Vauban built them in the early eighteenth century, and the model of 1706 does not differ materially from what we see on the ground today. More often, the models show fortified traces that have neither entirely disappeared nor survived intact but have been absorbed into the pattern of modern cities, in many cases continuing to determine the line of their street systems.

There are examples not only of the work of the French school of engineers but also of foreign schools. The model of Ostend, made in 1708 (Figure 5.18), thus shows the bastioned trace built by William of Orange at the end of the sixteenth century to protect this model medieval "new town." As we have seen, many of the models include parts of the fortifications built by Spanish engineers of the sixteenth and seventeenth centuries. At Berg-op-Zoom, shown in a model of 1750, the fortifications were the work of the Dutch engineer Coehorn, contemporary of Vauban. The French captured the town in 1747 and had time to make the model, which perhaps helped them when they recaptured it in 1795. In 1814 Berg-op-Zoom was returned to the United Provinces and is now part of North Brabant. The models, then, give us a fine conspectus of the work of most of the leading fortification engineers of early modern Europe; in some cases they also allow us to reconstruct vanished works.

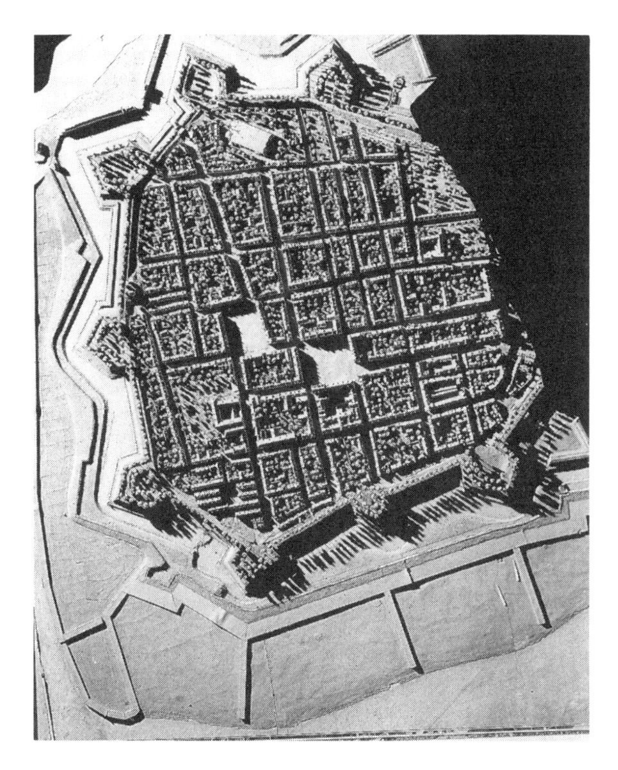

FIGURE 5.17. Vertical view of the model of Nieuport (c. 1698)

FIGURE 5.18. Vertical view of the model of Ostend (c. 1700)

CONCLUSION

AS A SOURCE, these models have some disadvantages, the greatest of which is that is difficult to display them in a way that allows them to be fully appreciated. The *Grande Galerie* (though not designed for this purpose) was a marvelous space for the early models, and some are most advantageously displayed in the Musée des Beaux-Arts at Lille, where they can be fully appreciated in a huge hall that allows the visitor not only to see them close up, but also to go to an upper floor for an aerial view of them. However, this is hard to arrange in most spaces.[54]

Against these disadvantages must be set many merits. The model is the least abstract of all representational types and so accessible at first glance to the untrained eye. Even seasoned map readers are often unable at once to seize the significance of all the contour lines on a map, but with a model every variation of the relief can be seen at once. Since the model by definition aims at a copy of surface appearance as faithful as possible, it takes in by chance many features that were not of primary interest to its creators. The engineers of Louis XIV, for instance, were not interested in showing the olive trees around Saint-Tropez, except in so far as they helped or hindered military operations. But the fact that they put them and many other agricultural features into their models is now of extraordinary interest to the historian.

As well as being a comprehensive representation of surface reality, the model is also an object that can be seen from many different angles. Whereas a town delineated on paper has to be seen either in profile, as a bird's-eye view, or as a plan, a town shown as a model can be visualized in all these ways by the viewer, simply by altering the angle of vision. Since each mode of paper delineation has its uses, it is no small advantage to be able to combine them all into one representational form.

This advantage was well understood by many commanders, from Philip II onward. We have the testimony of Napoleon that "there is no better map than these relief models," and Eisenhower made much the same observation in 1944. When a large body of people needs to be rapidly acquainted with the outlines of a city, there is no substitute for a model of it, as many tourist centers of our time have come to realize. In short, to portray the city it is often best to model it.

Notes

1. The best recent summary of the history of these models is by Antoine de Roux, Nicholas Faucherre, and Guillaume Monsaingeon, *Les Plans en relief des places du roy* (Paris, 1989).

2. This emerged clearly from an international conference held at the Invalides in 1990 on the theme of city models; see the *Actes du Colloque International sur les plans-reliefs au passé et au présent*, ed. André Corvisier (Paris, 1993), hereafter cited as *Actes du Colloque*.

3. See Franz Bischoff, "Les maquettes d'architecture," in *Les Bâtisseurs des cathédrales gothiques*, ed. Roland Recht (Strasbourg, 1989), 286–95; also *Filarete's Treatise on Architecture*, trans. and ed. John E. Spencer (New Haven, Conn., and London, 1965), 16.

4. Sir John Hale, *Renaissance War Studies* (London, 1983), 74.

5. Roux, Faucherre, and Monsaingeon, *Les Plans en relief*, 18.

6. Ibid., 52.

7. Gerald Strauss, *Sixteenth-Century Germany: Its Topography and Topographers* (Madison, 1958), 73.

8. Roux, Faucherre, and Monsaingeon, *Les Plans en relief*, 18.

9. Simon Pepper and Nicholas Adams, *Firearms and Fortifications* (Chicago, 1986), 193.

10. See the summary by Carlo Gottardi, *Actes du Colloque*, 175–79.

11. According to Amelio Fara, in his talk at the 1990 conference.

12. See *Augsburg: Geschichte in Bilddokumenten* (Munich, 1976); Fritz Schnelbögl, *Dokumente zur Nürnberger Kartografie* (Nuremberg, 1966), 75; and Uta Lindgren, *Actes du Colloque*, 167–74.

13. See the introduction to Louis Grodecki, *Plans-Reliefs des villes Belges* (Bruxelles, 1965).

14. James S. Ackermann, *Distance Points: Essays in Theory and Renaissance Art and Architecture* (Boston, 1991), 373; and Henry Millon, "I modelli architectonici nel Rinascimento," in *Renascimento da Brunelleschi a Michelangelo*, ed. Henry Millon and Vittorio Magnago, (Milan, 1994), 19–73.

15. Pepper and Adams, *Firearms and Fortifications*, 60.

16. Richard Kagan, ed., *Spanish cities of the Golden Age* (Berkeley/London, 1969), 49.

17. Cited in Eduardo de Mariategui, *El Capitan Cristóbal de Rojas* (Madrid, 1880), chap. 6.

18. Juan Miguel Munoz Corbalan, *Actes du Colloque*, 181–94; for a photograph, see José Antonio Calderon Quijano, *Las Defensas del Golfo de Cadiz* (Seville, 1976).

19. See Hale, *Renaissance War Studies*, 227.

20. According to Charles van Heuvel, *Actes du Colloque*, 57–70.

21. See C. J. Zandvliet, "Twee Zeventiede-Eewse Collecties Maquettes van Nederlandse Vestingen en Versterkte Steden," *Jaarboek* 1990/91 of the Stichtung Menno van Coehoorn.

22. See Claire Lemoine-Isabeau, *Actes du Colloque*, 195–202.

23. A. Charbonneau, *Le Plan en relief de Québec* (Quebec, 1981).

24. See Ernst-Heinrich Schmidt, *Actes du Colloque*, 143–53.

25. See Vice-Admiral Paris, *Notice du Plan en relief du canal maritime de Suez* (Paris, 1882).

26. See Jean Langenberger, *Actes du Colloque*, 155–66.

27. See Leonard Abrams, *Our Secret Little War* (Bethesda, Md., 1991) and official manuals such as *The Construction of Topographical Models and Relief Maps*, published by the Army Map Service of the Corps of Engineers (Washington, D.C., 1944).

28. International Colloquium on Progress in Terrain Modelling, *Progress in Terrain Modelling* (Denmark, 1987).

29. See W. W. Ristow, ed., *Three-Dimensional Maps* (Washington, D.C., 1964).

30. Cited in Martha Pollak, ed., *Military Architecture, Cartography, and the Representation of the Early Modern European City* (Chicago, 1991), 65. Maggi's contemporary in the service of the king of France, Nicholas de Nicolay, also claims to have made "a very curious relief model" of Constantinople during the 1550s; see Robert Karrow, *Mapmakers of the Sixteenth Century and Their Maps* (Chicago, 1993), 437.

31. Pollak, *Military Architecture,* 67.

32. Roux, Faucherre, and Monsaingeon, *Les Plans en relief,* 22.

33. Catherine Brisac, *Le Musée des Plans-Reliefs* (Paris, 1981), 9.

34. Ibid., 10, and Alain Boulaire, *Actes du Colloque,* 115–22.

35. Brisac, *Le Musée,* 36.

36. Roux, Faucherre, and Monsaingeon, *Les Plans en relief,* 41.

37. Ibid., 51.

38. Grodecki, *Plans-Reliefs des villes Belges,* 11.

39. Roux, Faucherre, and Monsaingeon, *Les Plans en relief,* 49.

40. Brisac, *Le Musée,* 39.

41. Roux, Faucherre, and Monsaingeon, *Les Plans en relief,* 43.

42. Brisac, *Le Musée,* 18.

43. Ibid., 19.

44. Cited by Fernand Beaucour, *Actes du Colloque,* 83–90.

45. Brisac, *Le Musée,* 24.

46. Grodecki, *Plans-Reliefs des villes Belges,* 14.

47. Brisac, *Le Musée,* 24.

48. Grodecki, *Plans-Reliefs des villes Belges,* 14 and Brisac, *Le Musée,* 26, offer differing accounts.

49. See the contribution of Antoine de Roux to the 1991 conference, "Les Plans en relief, une source capitale d'information pour l'histoire des villes et de l'équipement des campagnes," *Actes du Colloque,* 23–32.

50. Roux, Faucherre, and Monsaingeon, *Les Plans en relief,* 92.

51. See the accounts in both Brisac, *La Musée,* and Roux, Faucherre, and Monsaingeon.

52. Grodecki, *Plans-Reliefs des villes Belges,* 15; when he wrote, in 1965, there was a plan to transfer the collection to the Petite Ecurie du Roi at Versailles. (The move did not occur.)

53. See, e.g., Roux, Faucherre, and Monsaingeon, *Les Plans en relief,* 119.

54. See the publication of the Musée des Beaux-Arts de Lille, *Plans en relief: Villes fortes des anciens Pays-Bas Français au XVIIIe* (Lille, 1989).

SIX

The *Plan of Chicago* by Daniel H. Burnham and Edward H. Bennett: Cartographic and Historical Perspectives

Gerald A. Danzer

PEOPLE HAVE, from the very inception of urban life, made maps of their city because they could not see it the way they wanted to see it.[1] Perhaps there was no promontory nearby to afford a panoramic view of the entire town. Yet they wanted to come to terms with the city, to see it or envisage it in its totality. A concept of the metropolis as an organic whole was a prerequisite for urbanity, and the making of maps was one way to address this need for a comprehensive depiction. In some situations, perhaps, a convenient viewing place might afford an expansive view, but the vista was often not as grand or impressive as might be desired. Some major streets or buildings might be obscured by the natural lay of the land or pushed out of sight by the spread of the settlement. Thus even when standing on a lofty overlook, a viewer needed either a mental map or a pictorial image at hand to fully comprehend the city in its entirety.

The development of urban cartography over the centuries produced a variety of methods and techniques for helping residents or visitors perceive the shape and extent of towns and cities. The definition of a map naturally expanded to include views as well, leading to a variety of cartographic images ranging from abstract plans to realistic views.[2] As in all maps, the key elements for producing satisfactory urban images were selectivity, simplification, and the use of symbols. For many purposes a simple diagram of the major streets could suffice, but in that case the vertical dimension of the urban fabric had to be omitted. Urban cartographers thus turned to profiles and perspective views to provide a fuller portrayal of the built environment. One of the richest chapters in the history of cartography tells the story of the development of urban views after the introduction of the printing press into Europe. It begins in 1500 with the remarkable woodcut of Venice on six sheets by Jacopo de' Barbari[3] and proceeds to the rich volumes of the *Civitates Orbis Terrarum* produced by Georg Braun and Franz Hogenberg beginning in 1578.[4] The leading cities of the New World were included in this collection; every major English city in North America had received a series of cartographic portraits by the end of the colonial period.[5] As towns and cities spread across the continent, they engendered a myriad of urban images, leading up to such elaborate productions as Camille N. Dry's panoramic view of St. Louis in 1875 which took 110 plates to complete a nine-by-twenty-four-foot image that literally pictured every residence in the city.[6]

In large measure, the purposes of these elaborate pictorial maps of the city were rooted in local pride, the development of group consciousness, and the promotion of civic loyalty. They served as symbols of urban achievement: almost every modern city in the Western world possesses a series of printed views to mark the stages of its development. These representations of the city's topography were accompanied by a host of other types of urban maps tailored to meet the particular needs of travelers, real estate interests, sanitary engineers, insurance underwriters, social scientists, specialists in health and recreation, and city planners.[7]

Relating the various representations of a city to each other and connecting them to the particular place as it developed over time are among the leading contributions that a historian of cartography has to offer to urban studies generally and to the modern city in particular. To understand a twentieth-century metropolis in cartographic terms, one needs more than one map. In particular, one must investigate the interrelationships between abstract maps, bird's-eye views, and other representations. The formal plans for cities put forward by various agencies for twentieth-century cities provide useful case studies to make this point. Thus a series of studies looking at cities through their maps might well conclude by looking at perhaps the most celebrated plan of the City Beautiful movement in the early twentieth century.

Typical historical questions will launch this study of the Plan of Chicago: What circumstances led to its publication? Who was Daniel Burnham? Does his career as an architect provide links to his role as a planner? But when the investigation focuses more narrowly on the plan itself, the cartographic lens brings several specific concerns about maps clearly into view: the dynamic between land and water, for example, or the relationship between the natural setting and cultural features. Finally, the two analytical approaches—from history and cartography—will merge in considering the book as an atlas. What maps were selected for inclusion? How are they connected to each other? What is the significance of their particular sequence? How do images and text relate to each other? What is the heart of the plan itself? How does the focal point of the proposed metropolis reveal the power of the center in its cartography? And, finally, what meanings of a more general character might be teased from a close look at the *Plan of Chicago* as a collection of maps?

Plan of Chicago, 1909

In 1906, when Daniel Burnham started to work on a plan for Chicago, a city he had known for half a century, he directed his staff to collect every type of urban map and view they could assemble. To comprehend a city it was, quite obviously, necessary to study as many different images as possible. But even a choice assortment of books, maps, and plans could not substitute for a high vantage point affording vistas of the city itself. Therefore, as soon as Burnham was commissioned by a group of businessmen to start work on a comprehensive plan, he had a penthouse office constructed on the top of a tall building in the center of the city to house his staff.[8]

Above the seventeenth story of the newly constructed Railway Exchange Building, directly across Michigan Avenue from the Art Institute, the architect and his colleagues could look out on the city and ponder its character, assessing its needs from an Olympian height, searching for solutions to its problems and for ways to realize its potential. As they searched and pondered they soon had access to a whole library of materials on the form of cities and the concepts of urban planning. Some of these documents were already in Burnham's office, in the same building, one story down, as a result of his pioneering efforts in producing major plans for Washington, D.C., Cleveland, San Francisco, and Manila.[9] Because his architectural firm had executed commissions in many other cities, he had many contacts to draw on for up-to-date information on developments in other centers such as Boston, New York, and Philadelphia. Moreover, recent travels in Europe had introduced him to the latest examples of European plans and projects. He extended these contacts by requesting the Department of State through its consular service to furnish "valuable information relative to civic developments now in progress" throughout Europe.[10]

The *Plan of Chicago Prepared under the Direction of the Commercial Club during the Years MCMVI, MCMVII, and MCMVIII* by Daniel H. Burnham and Edward H. Bennett, edited by Charles Moore, a corresponding member of the American Institute of Architects, was published in Chicago by the Commercial Club in 1909. It was sumptuously produced for a list of subscribers who had pledged funds to support the project as a whole.[11] The book itself, like the plan it presented, was a private effort done in grand style by and for an elite group. It was a composite volume combining a text drafted by Burnham and completed by Moore with a host of illustrations, maps, diagrams, and plans assembled under the direction of Burnham's assistant, Edward H. Bennett. "Legal Aspects of the Plan of Chicago" by Walter L. Fisher, counsel for the Plan Committee of the Commercial Club, appeared as an appendix in the printed volume.

The illustrations appearing in the book, over 150 in number, command especial interest. They were carefully selected from a variety of sources or specifically prepared for the project.[12] The intention from the very beginning of the undertaking was to present the plan as an exhibit at the Art Institute. The book would then utilize the same materials, presenting, as it were, an expanded traveling exhibit along with the text of the proposal. As it turned out, the book's illustrations and text became quite independent of each other. They are coordinated to be sure, but many of the pictorial elements are not directly referred to in the text and vice versa.[13] A reader almost needed to go through the book three times to really understand it. The first reading would focus on the pictures and maps to get an overall impression of the proposal. The second reading would follow the argument of the text itself. The third reading would then relate the text to the plates and presumably to the city itself.

Edward H. Bennett ranks as a coauthor of the plan in large measure because of his work in drawing maps or plans, selecting illustrations, and supervising the paintings, drawings, and maps prepared by the staff. Two notable artists were employed by Burnham to provide renderings that would add to the aesthetic appeal of the plan. Jules Guerin, working in New York, painted a series of large views that formed the core of the exhibition. These were then reproduced in color in the book. Fernand Janin, a young French artist and Bennett's friend, came from Paris to sketch several maps and views, especially dreamy images of the proposed civic center at the heart of the proposal. In addition, five other individuals contributed drawings or illustrations for the book.

Janin's impressive sketches, Guerin's expansive paintings, and Bennett's maps have come to be identified as the plan itself. The text, in historical perspective, has faded into the background as a document of the City Beautiful movement. The visual materials, however, have endured as powerful images of an ideal city, taking on a life of their own. In retrospect, therefore, it is more than a courtesy to refer to the book under multiple authorship, although in popular usage the effort is commonly called the "Burnham Plan" after the senior member of the team.

The *Plan of Chicago* may be profitably studied as a treasured book, as an artifact of striking artistic merit, and as an example of the very best printing techniques of the early twentieth century.[14] Although the volume shows evidence of haste in its final production, it was printed by the Lakeside Press in Chicago whose owners, members of the Commercial Club, spared no effort to make the volume an elegant example of the printer's art. The quality of the illustrations, color plates, paper, layout, and various embellishments made the publication a landmark in the history of printing in the United States. It was one of the first lavishly illustrated books to make extensive use of color plates.

In some ways the *Plan of Chicago* could be classified as an exhibition catalog. The original paintings, drawings, and maps were exhibited from time to time at regular meetings of the various committees formed by the Commercial Club to oversee various aspects of the developing plan. Burnham would also take small groups of businessmen back to his office to see the latest Guerin painting or Bennett map as a convincing way of pushing his ideas. When the book was officially published on July 4, 1909, much of the original art work was placed on exhibit at the Art Institute. Ben Holden, one of the staff members who developed the plan, then took selected renderings to London and to Düsseldorf where they were exhibited as examples of the latest ideas in American urban planning.

The *Plan of Chicago* could also be considered an atlas, a bound collection of maps of the city accompanied by a parallel text and other supporting elements of a pictorial nature. With an orientation to the future rather than to the present or the past, the authors turned to several traditions in urban cartography. An urban plan basically is a scale drawing laying out the streets and public places of a town or city. It may outline the major geographic features, but its chief purpose is to divide the urban land into public ways and private spaces, into streets and blocks in Chicago terminology.

The plan of a city often precedes the actual development of a place. Thus the plan seeks to overcome a limitation in time just as the bird's-eye view seeks to overcome a limitation in space. At the very beginning Chicago needed a plan to tell its citizens what the town would look like when the streets were laid out. The earliest published maps of Chicago are indeed plans that portray a town that would be built in the future rather than one that already existed. Pioneer buildings, gardens, paths, and trails had to be relocated to fit into the new design when the plan moved from paper to surveyor's stakes on the ground. The first maps for an American city are often abstract, tentative indicators of change, oriented to the future.

Any cartographic image requires readers to bring several contexts to the map to give it meaning. To read a particular urban plan a person needs a general idea of settlement patterns and building styles, economic functions, traffic flows, and perhaps even some familiarity with the local landscape. A reader, as it were, has to put flesh on the skeleton of a plan's layout of streets, plazas, and parks to make it a living, meaningful document. An atlas assembling a variety of pictorial resources would help develop this conceptual matrix.[15]

Burnham and Bennett, when they stood on the roof of the Railway Exchange Building, quickly concluded that they needed more than a map to help the citizenry envisage a new Chicago. The urban mosaic spread out before them in orderly fashion, but it was drab, crowded, and ugly, filled with noise and smoke. It extended back from Lake Michigan in every direction and, as any businessman would attest, its influence reached across and around the lake as well. Their task, as Burnham and Bennett envisioned it, was to produce an atlas that would convince citizens who already knew the city of 1909 at the functional level that it could become something else. The planning team needed images of the city that would provide a sense of large patterns and future possibilities. Their approach was to take a basic vision of what Chicago might be and illustrate it with a variety of maps, views, plans, diagrams, sketches, drawings, paintings, and word pictures.

One master map would not do. The same subject should be portrayed in a variety of media, pictured from various perspectives, at different seasons of the year, at daybreak and at sunset, in black and white, and in color. The reiteration of the same vision in many formats would make it more familiar, more natural, more acceptable, and therefore more possible. By emphasizing basic themes again and again in a plethora of pictorial images, the planners took care to show that these elements were also common denominators found in the great cities of Europe and in contemporary blueprints for progressive American cities. They provided pictures of other places, especially of Paris as transformed under Baron Haussmann, to build an image of what a great city should look like. Then they modeled a future Chicago in Parisian dress.[16]

This account undoubtedly simplifies the picture too much, and it must be balanced by a careful reading of the text and a judicious appraisal of individual maps, but it provides the core of an explanation for the great reputation of the Burnham project in the history of American urban planning and its continuing appeal to the citizens of Chicago. Burnham's often quoted words, "Make no little plans; they have no magic to stir men's blood and probably themselves will not be realized," do not appear in the *Plan of Chicago*, although the entire project may be said to reflect that motto which his first biographer claimed was formulated in 1907, in the midst of his planning effort.[17] "Make big plans," the passage continues, "aim high in hope and work, remembering that a noble, logical diagram once recorded will never die, but long after we are gone will be a living thing, asserting itself with ever growing insistency."

Given these terms as a measure of success, the 1909 plan must be accounted as a major achievement. Scenes from the plan are still painted on Chicago buildings eighty years later in celebration of the city's vitality. Memory of the plan is kept alive in such objects as a huge twenty-two-by-eighteen-foot tapestry adorning the lobby of a postmodern skyscraper that reproduces plate 132 of the plan.[18] Guerin's westward view of the proposed civic center, at dusk, just after a summer shower has made the plaza glisten, remains a living symbol of Chicago's aspirations.

But Burnham's credo continues, "Remember that our sons and grandsons are going to do things that would stagger us. Let your watchword be order and your beacon beauty." A plan for a city's development also needed to be open-ended to let the urban fabric accommodate the achievements of future generations. Thus while making big plans and noble diagrams, planners should keep things in order and strive for beauty. Indeed, under close inspection the 1909 vision rigidly adheres to the grid for the city established by the congressional land survey and the original plan laid out by the pioneer surveyor, James Thompson, in 1830. Burnham's bold advice about big plans at the beginning thus seems to be qualified in each successive sentence, ending in a plea for discipline and grace.

Plans are, of course, produced by individual planners, and the angles of vision provided by their life experiences go a long way in helping readers to understand the designs. This is also true for maps, and urban plans, which look to the future, are, by their very nature, much more transparent in the way they illuminate the outlook of their creators. Thus if one is to develop an understanding of the *Plan of Chicago* as an atlas, it is helpful to start with a brief biographical sketch of the central figure.

Daniel Burnham

Daniel Burnham carried a name that reached back to the early period of English settlement in New England, his ancestors arriving in Ipswich, Massachusetts, in 1635.[19] His branch of the family moved to Connecticut, then to Vermont, and, finally, in 1811 to the New York frontier.

The senior author of the *Plan of Chicago* was born on September 4, 1846, in Henderson, New York, a small town near the northeast corner of Lake Ontario. His mother traced her ancestry back to the Pilgrims and brought a dedication to Swedenborgian Christianity to the Burnham household of six children, of whom Daniel was fifth in line. After a sequence of moves and business ventures that met with little success, Edwin Burnham, at his wife's insistence, moved his family to Chicago where Edwin's brother had established a successful law practice.

Thus Daniel Burnham came to Chicago with his parents in 1855, the very year that a decision was made to remove the last vestiges of Fort Dearborn, signaling that the city had left its pioneer period behind. Large railroad buildings would soon be erected on the site of the old military reservation. "During the last six months," the *Democratic Press* reported on February 12, 1856, "the architects have been working incessantly, drawing up plans for new and splendid buildings; the brickmakers at all points in Illinois and Wisconsin have contracts to the utmost extent of their ability to fulfil, while the cry is yet for more; far out on the prairies, north and south and west, spacious and substantial dwellings are springing up; and upon the east old Lake Michigan is daily encroached upon by the Illinois Central Railroad Company, who are rearing immense piles of stone and mortar" for its new depots and warehouses.[20]

Daniel Burnham was nine years old when he arrived in Chicago, a very impressionable age. One cannot but wonder at the impact the city made on the young boy. It is interesting to note how the newspaper account reflects many of the themes that would reappear in the *Plan of Chicago:* the changing city, the guidance provided by architects, the connections between Chicago and the midwestern states, the spreading of the metropolis across the prairies, the refashioning of the lakefront, and the cry for "more."

A few weeks after the newspaper report was printed, James T. Palmatary, an artist, arrived in Chicago to begin work on a large lithograph view of Chicago that has become, in retrospect, one of the most celebrated portrayals of the city.[21] Palmatary did his sketches throughout the city in 1856, and his print appeared, in two large sheets, in 1857. The cityscape artist caught the city just as railroad cars, canal barges, and lake boats were making it the hub of the Middle West. He pictured the Illinois Central terminal buildings as completed structures although, at the time he sketched the site, the bricks for the buildings were just arriving from distant brickmakers in Illinois and Wisconsin.

As a schoolboy Daniel Burnham was notorious for spending his time drawing and sketching instead of tending to his studies. Did he watch Palmatary make his sketches on the city streets? Did he pore over the magnificent scene when the artist's cityscape appeared in print? His mother had insisted that the family come to Chicago for the educational and cultural advantages it offered. Did she take her young son with an interest in drawing to see Palmatary's view when proof copies were placed on public display? We do not know the answer to these questions. But we can rephrase the questions to make them less biographical and more relevant to the concerns of urban cartography: What do Palmatary's "View of Chicago" in 1857 and Burnham and Bennett's *Plan of Chicago* in 1909 have in common? In what respects do they differ? What changes occurred in the half century that separated them in time? How did each image help its audience grasp the idea of a city, its purposes and its life, its form and structure? Can one understand any city better by studying these particular examples of the urban idea?

Both Palmatary's print and the *Plan of Chicago* tried to capture the essence of the city for different generations. One attempted to portray the whole city as it existed in 1856 in a single view. The other wanted to show Chicagoans what their city could become if some basic principles of planning were followed. Palmatary, a visitor, celebrated how far the city had come. Burnham, reflecting a lifetime witnessing Chicago's incessant expansion, looked to a future even more glorious.

The Burnham family participated in the prosperity brought on by the continued expansion of the city, rising over time into comfortable status. In a few years, young Daniel was sent back to Massachusetts for preparatory school, but he failed to gain admission to either Harvard or Yale. Some commentators have pointed to this absence of a college education as a key to the architect's later reverence for academic authority.[22] Returning to Chicago, the young Burnham tried his hand at business, then did some work in William Le Baron Jenney's architectural office, but soon left the city to seek success in a Nevada mining district. He tried both prospecting and politics, failed at both, and returned to Chicago to pick up where he left off, pursuing a career in architecture. Before he left on his Western adventure he had written to his mother, "I shall try to become the greatest architect in the city or country. . . . There needs but one thing. A determined and persistent effort."[23] It was now time to begin.

Burnham and Root

When Burnham returned to Chicago from his adventure in Nevada he was twenty-four years old. While taking stock of his situation he worked briefly with several architectural firms. In each of these he failed to establish himself. In October, 1871, the Chicago Fire created a great need for architects to guide the rebuilding of the city. Shortly after the reconstruction started, Burnham took a job with the young architect Peter Wight.[24] In fact, Wight's firm had only one major commission in the downtown area. Instead, it picked up the slack in residential design while the more established firms concentrated on the large commercial projects downtown.

Wight apparently spent some time instructing his assistant in the practice of architecture. Burnham also undoubtedly acquired a lot of insight from the chief draftsman in Wight's firm, John Wellborn Root.[25] Root, born in Georgia in 1850, had spent the Civil War years in England where his family's well-to-do status enabled him to receive the very best education.

At the end of the Civil War, Root sailed to New York where he took a degree in engineering in 1869. After briefly working with several leading New York architects, Root decided to take advantage of the post-Fire opportunities in Chicago. He arrived in Wight's office at the same time as Burnham. The contrast between the two newcomers in background, education, and temperament did not stand in the way of their growing friendship. Then, in 1873, Burnham and Root decided to go into business for themselves.

The opportunities presented by the rebuilding of the city were set back by the Panic of 1873, but the two young architects struggled along designing a variety of small houses and barns. Their big break came in 1874 when a friend secured for them the chance to design a large house for John B. Sherman, the manager of the Chicago Stock Yards. Burnham more than took advantage of the opportunity, securing the hand of Sherman's daughter in marriage as well.

Burnham and Root were soon fashioning residences for Chicago's prominent citizens, and these contacts led to opportunities to do a variety of other buildings. Their first chance to design a downtown structure came in 1881 with the seven-story Grannis Block on Dearborn Street. The following year Burnham and Root secured a place in architectural history with their design for the Montauk Block on Monroe Street.[26] It was an epoch-making building on several counts. At ten stories it was the first office building in Chicago to tower above its neighbors. Pushed by the Boston investors to keep down the costs and to build as much floor space as possible, Burnham and Root developed a novel type of foundation to sustain the weight of a tall building on Chicago's soft soil. They also stripped the structure of unnecessary ornament, used the newest technology wherever possible, provided for electricity, and used a central shaft to carry the utilities to each floor. They also pushed the building up to ten stories from the planned eight. Perhaps the first office building to be called a skyscraper, the structure attracted the attention of architects, builders, and investors across the nation. It also set the course for the future development of Chicago's downtown.

Burnham and Root continued to fashion a variety of residential, commercial, industrial, and institutional buildings.[27] Their firm never became exclusively devoted to tall commercial structures, but it is remembered today primarily for its large office buildings. The monumental Rookery Building (1885–88) reached eleven stories and occupied a very large site. It was commissioned by the same Boston investors, less cost conscious in this case because they aimed to create an up-to-date luxury structure. The interior space was hollowed out to form a large court with four office wings grouped around it. The interior yard was then covered with a glass dome to enclose a two-story shopping center at the base of the building.

Other pathbreaking architectural firms were erecting landmark buildings in Chicago's central business district at this time. Perhaps the concentration of innovative talent thrived on the competition. Burnham and Root, however, enjoyed advantageous connections with some leaders of Chicago society as well as with the Boston investors, Peter and Shepard Brooks, who worked through a talented agent in Chicago, Owen Aldis.

The Monadnock Building, a series of connected structures erected between 1889 and 1891 on a narrow lot at Jackson Boulevard and Dearborn Street, represented the epitome of the collaborative efforts mounted by Boston finance and Chicago's architectural innovation. Returning to the no-frills, businesslike approach used in the Montauk, the architects raised the building height to sixteen stories. At first, they stayed with the traditional wall-bearing masonry architecture instead of utilizing the new skeletal structure of ironwork that William Le Baron Jenney was using for his sixteen-story Manhattan Building being erected down the street at the same time. In effect, to pay for the thick walls and spreading foundations required by the Monadnock, Burnham and Root were forced to drop all the exterior decoration. The result, in hindsight, was to usher in the twentieth century a decade before its time.

The Monadnock Building was immediately recognized as a special, innovative structure. It filled with tenants immediately and remains a viable office building today. Moreover, it defined a new relationship between the architect and the city, between the artist and society. Walter Behrendt, a critic writing in the 1930s, said it best:

> In its rigid functionalism, demonstrating a new conception, it became a landmark of modern building: the architect, as an artistic personality, steps back behind the commission given him by society. In an act of self-denial, he puts his individual forces into service for common needs, arising from the new social evolution. In this attitude is manifested the truth that building is a social art. The consequence of this conception . . . is to bring building again into a reasonable organic relation to the actual social and economic world, thereby re-establishing that indispensable identity between the content and the form of life, which is missed in the works of those who have turned their backs on their time.[28]

It would be a simple story if the tradition of modern American urban planning, fathered by Daniel Burnham, came out of the new conception of architecture heralded by the new skyscraper on Dearborn Street. But it did not. Like its namesake, the Monadnock stood as a lone peak while the new skeletal style of construction and contemporary aesthetic taste pulled the architecture of office buildings in new directions. Burnham and Root continued to follow their natural inclinations to make more of an artistic statement with their designs. Yet they were so much a part of the social and economic order that the creative tension between business and aesthetics, between order and beauty, could be used profitably to analyze their later work, and especially to comprehend Burnham's *Plan of Chicago.*

This stretching in two directions, toward the functional and the aesthetic, is illustrated in two of Burnham and Root's last major skyscrapers. Both were named temples, both were located in Chicago, and both were completed in 1892, the year following Root's death.

The Women's Temple, commissioned by the Women's Christian Temperance Union (WCTU), followed the shape of the capital letter "H" to provide ample light and ventilation for offices in a large structure. Reaching thirteen stories in height, its utilitarian design was largely concealed by an elaborate three-story roof that used a steeply pitched elevation to display French Gothic decorations. Leaders of the WCTU liked the building, but it soon became economically obsolete and was demolished in 1926. Later architectural critics and historians have largely dismissed the building as a failure, "a tall but ultimately fussy attempt to make a skyscraper seem 'feminine.'"[29] Yet it might be one of the structures whose loss the citizens of the later city will mourn the most. It represented the city of commercial and industrial strength well with its strong rounded arches at the base and a Romanesque cornice enclosing the business block. The shoulders of the office building supported a jeweled reliquary, order sustaining beauty, a gem on the prairie.

The second "temple," this time for the Masonic Order, was conceived originally as the Capital Building. It was consciously designed to be the world's tallest office building. A commercial-style office block of fourteen stories was placed on top of a compressed two-story base that used the usual sequence of Romanesque arches to visually support the massive masonry above it. In reality, of course, a steel frame supported the entire structure. Above the two cornice stories that marked the termination of the office block, a steep gabled roof added another three stories to the building's imposing height. Root, as a designer, was said to be uncomfortable with the elevation of the structure. Its height went way beyond the proportions set by aesthetic conventions. It was too tall, towering beyond the limits of civility, even if crowned by a massive roof elevation to weigh down the ungainly stacks of offices and fasten them to the street below.

Root's sudden death came at a crucial moment. In the autumn of 1890 Burnham and Root had been retained as consultants by the World's Columbian Exposition. A group of outstanding American architects had been assembled by the board to design individual buildings for the coming world's fair. It was at the very time when this distinguished group was meeting in Chicago to make final decisions on the buildings for the fair that Root passed away.

After Root's death, designers such as Charles B. Atwood, Frederick P. Dinkelberg, Dwight Perkins, Ernest R. Graham, and Pierce Anderson continued to work with Burnham, producing a whole catalog of noteworthy buildings: the Reliance, Fisher, and Railway Exchange buildings, the Marshall Field's store on State Street, the Flatiron Building in New York, Union Station in Washington, D.C., the John Wanamaker Department Store in Philadelphia, and rank upon rank of other large commercial buildings from coast to coast.[30]

When Daniel Burnham became a city planner he was already one of the most notable and influential architects in the country. In some ways his contributions to the team of Burnham and Root were those of the planning process: the ability to see a project as a whole and as part of a larger picture; to communicate the essence of a proposal to a client; to maintain the integrity of a scheme in the face of inevitable changes; to keep a variety of details in order; and to add those embellishments to a project that would elevate it from the mundane. Typically Burnham would come back to the office and show Root a sketch he had drawn for a potential client, size up the various aspects of the proposal, and then turn the project over to his partner to produce an appropriate design.

Each individual was gifted at his task. Once the design was completed, Burnham again took over the tasks of client relationships, supervising construction, and keeping the project on schedule and within the budget. When Root died and left Burnham alone with the coordination of the World's Columbian Exposition, a master plan had already been conceived, in this case by the outstanding figure in American landscape architecture, Frederick Law Olmsted. The strengths Burnham brought to the new job with the world's fair were precisely those that were most needed at the time.

The French Building at the fair featured a display on the "City of Paris" which exhibited forty paintings by French artists. The net result of these urban views was to celebrate the city as a work of art, much in the way that the carefully planned exposition grounds in 1893 demonstrated the potential for re-creating American cities. One impetus to the urban planning movement in the United States came directly from the Columbian Exposition experience. And Daniel Burnham was in the center of things.

At first it was the classical style of the fair buildings themselves that attracted attention. Then, with the approaching centennial of the national capital as the seat of government, an opportunity arose to improve the Washington park system. This led to the appointment of a commission, with Burnham as chair, which began its study by focusing attention on the relationship between parks and public buildings. The next step was to prepare a plan for Washington's mall and associated public spaces and buildings west from the Capitol. The Senate Park Commission's proposals, presented in 1901 through a series of reports, maps, paintings, and exhibits, led to the redesign of Washington which returned the city's development to the original baroque concepts of Pierre L'Enfant.

The plan for the national capital supplied a model for planning efforts in other cities.[31] One important aspect of the planning process was to furnish each presentation with painted views, suggestive drawings, and a variety of pictorial elements in addition to the usual maps and blueprints. The perspective renderings and bird's-eye views of the Washington plan were particularly effective. They were contributed by a number of artists, including Jules Guerin,[32] a former Chicagoan who had painted scenes of the Columbian Exposition and had made a successful career working in New York. The text of the published proposal for the capital city's parks was edited by Charles Moore,[33] a talented architect and man of letters who served as the commission's executive secretary.

A few years later, Daniel Burnham would call on Guerin and Moore once again to shape the public presentation of the plan for Chicago. Edward H. Bennett,[34] another key member of the planning team put together by "Uncle Dan," was a talented young architect recommended by Burnham's chief designer, Pierce Anderson. Early in 1903, when Burnham was busily engaged in a variety of projects, including a pathbreaking plan for a group of public buildings in Cleveland, he faced a deadline for a proposal for some new buildings at the United States Military Academy at West Point. Bennett was "loaned" to Burnham by a New York architectural firm to shape up the proposal. The entry failed to win the competition, but Burnham found Bennett's work so useful that the young architect was called to Chicago in 1904 to design some buildings for Chicago parks and, then, later that year, sent to San Francisco to work with Burnham on a comprehensive plan for that city. When the senior author sailed for the Philippines to develop plans for Manila and Baguio, Bennett was left in charge of the San Francisco plan which duly cited him as a coauthor. But the planning effort at the Golden Gate eventually would be frustrated by the course of events when earthquake, fire, and political infighting limited its impact.

Both men were back in Chicago in 1906 to begin work on the Chicago plan which would again recognize their collaboration as coauthors. Bennett would continue as a major figure in urban planning and civic architecture until his death in 1954. The imprint of his ideas on the built environment of both Chicago and Washington is strikingly evident today. So, in this respect, is the work of Jules Guerin, the artist who would become the paramount conveyor of the Chicago plan to the general public. His bird's-eye perspectives provided the most enduring presentations of Burnham and Bennett's plans. Guerin was called back from New York to Chicago on more than one occasion—to design the fire curtain of the Civic Opera, to provide murals for the city's largest bank and its temple of merchandizing, and to advise on the decoration of its great room in Union Station. Perhaps these later connections to Chicago helped to keep Burnham's vision for the city alive, but even more suggestive is the way the master was able to utilize the talents of each individual member of the team he assembled in 1906 to plan the transformation of the city on the shores of Lake Michigan.

FIGURE 6.1. Chicago: Bird's-Eye View Showing the Location of the City on the Shores of Lake Michigan. Painting by Jules Guerin. (*Plan of Chicago,* plate 1)

THE CITY ON THE SHORES OF LAKE MICHIGAN

THE FRONTISPIECE (figure 6.1) for the *Plan of Chicago,* a watercolor by Jules Guerin, presents a satellite view of Lake Michigan with Chicago at its southwestern reach. Farms, fields, forests, and surrounding communities seem to fall in concentric rings around the urban hub. A quiet, subdued mood permeates the scene, set by the softness of the colors and the haziness of the atmosphere.

Lake Michigan seems to set up a cartographic conversation with the landscape of the Midwest. The soft voice from the monochromatic waters engages the busier pattern of the lands stretching south and west. The lake intrudes from the north pointing to a jewel. A careful observer can locate some details of the Great Lakes. Green Bay is set off by the Door Peninsula in Wisconsin. Beyond is a long bluish stroke suggesting Lake Superior. The character of Lake Michigan's shore seems to be more than a thin line. Instead it is a blurred region of gradual transition between land and water, emphasizing the way the elements are connected, two aspects of an organic whole. Lake ports suggested by terra-cotta pearls are strung along the shore, setting off the gem at the end of the necklace. Milwaukee, for some reason, is absent and the port of Green Bay seems too distant to appear as a city.

Chicago, the focal point for the map, dominates the central panel if the painting is divided into the nine equal rectangles often used in set design. Evanston, where Burnham made his home, is exactly in the center of the canvas. The metropolis extends along the lakeshore for miles, seemingly attracted to the waters. As it ventures inland, the city assumes an irregular shape, interlocking with the surrounding countryside.

The high elevation of the observer's position in space, presenting an oblique view of the earth's surface, dynamically portrays the engagement between the city and the lake. The altitude dictates a scale that is too small to delineate any internal features of the city, its river, streets, districts, sectors, or neighborhoods. The artist has suggested a rich texture in the urbanized region, but specific details seemingly would detract from the overall message. Chicago, like Lake Michigan, is monochromatic.

The terra-cotta color marking the city on the painting matches the color of the word "Chicago," prominently displayed on the title page which lies opposite the frontispiece. Thus, although the map seems to lack any lettering, a careful observer will note how the color coordination between the painting and the title announces a major theme of the presentation: the search for connections. This subtitle ploy is just one of many techniques used by the book's designers to tie its varied constituent parts together into a unified whole. An atlas is more than a collection of maps; it is a coordinated view of the earth.

The legend printed in small type below the frontispiece identifies the illustration as "number 1" and refers to the painting as simply "Chicago." A longer phrase then spells out the purpose of the graphic aid: "Bird's-eye view showing the location of the city on the shores of Lake Michigan, together with the smaller surrounding towns connected with Chicago by radiating arteries."

Actually more than the surrounding towns seem to be connected with the metropolis. The "radiating arteries" are faintly suggested here and there, reaching out to the various municipalities which are given the same color as the metropolis. But Chicago's reach embraces the fields and forests as well as the towns.

Light green patches suggesting fields of growing crops seem to be cut out of the darker green of forest lands which look like they formerly covered the landscape. The prairie is suggested only in long reaches in the top left portion of the image. This gives the impression of a serious geographical error in that most of the native vegetation in the Chicago region was tall-grass prairie with wooded areas appearing only as scattered groves or riverine strips. The use of artistic license here enhances the sense of interconnectedness between central place and hinterland, but it does so at the expense of geographical accuracy.

The tensions between accurate depiction and overall impression, between what a map records and what it conveys, between what it says and what it means are thus revealed at the very beginning of the *Plan of Chicago*. The advice of Burnham's credo seems to establish criteria for resolving these issues of cartographic design: "Make no little plans." "Aim high in hope." A "logical diagram . . . will be a living thing." "Let your watchword be order and your beacon beauty."

Lake Michigan established the starting place for visualizing Chicago, planning its future, and mapping its potential. "The lake is always east" is a popular expression residents tell visitors to help them get their bearings. As in Guerin's painting, the lake points to the site of Chicago, creates its facade, provides its orientation, and suggests an appropriate scale to set its goals.[35]

Lake and Grid

Lake Michigan figures prominently in the rest of the *Plan of Chicago*. It often provides the foreground for a picture, the background for a sweeping panoramic view, the first point of orientation on a map, or a baseline for architectural display. The lake as a whole, however, is the subject of only one other map, "Chicago, and Diagram of Lake Michigan" (figure 6.2). Its legend points to a thin orange line that traces the shoreline, the "proposed roadway to connect all the towns along the shores of the Lake." The map, however, seems to have been drawn to serve additional, even more fundamental, purposes. Although not given a full page in the book, the original piece, done in colored pencil, is quite large, measuring more than three by five feet. It had a prominent place in the exhibition at Düsseldorf in 1910 where an orientation to the American landscape was especially needed.

Unlike most of the other maps and views in the book, this one carefully outlines the exact shape of the lake, precisely locates and identifies

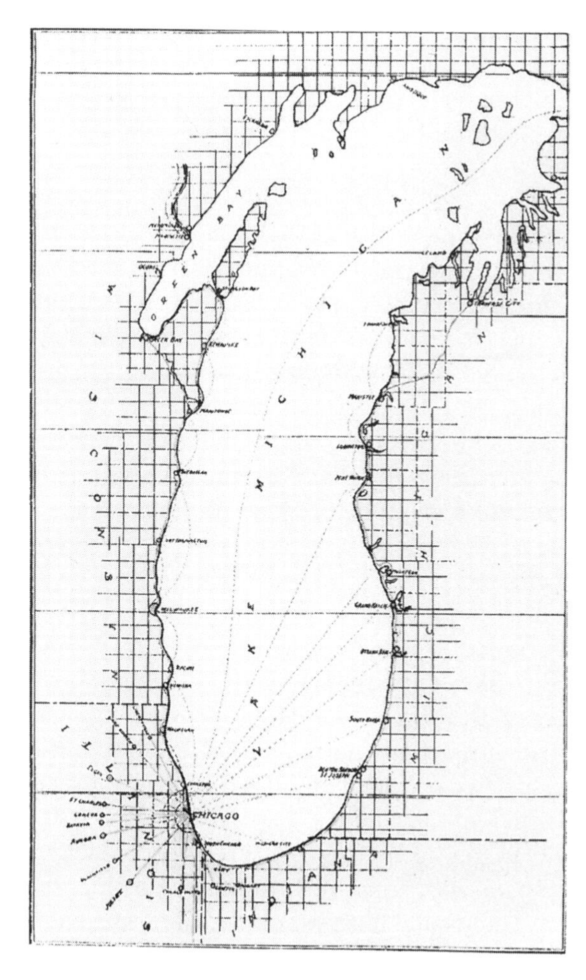

FIGURE 6.2. Chicago, and Diagram of Lake Michigan. (*Plan of Chicago*, plate 35)

several dozen cities and towns in the region, and is oriented according to the cardinal directions with north at the top. Shipping routes are lightly traced, all radiating from Chicago, so that there is no doubt that the city is the hub of the map, even if pushed, by the great reach of the lake, to the bottom of the sheet.

The most curious element on the map, however, is the systematic enclosure of the entire lake in a grid outlining the township and range system of land division.[36] These surveyors' townships, approximately six miles square, were the fundamental units used by the federal government to divide the public lands, assess their value, and locate tracts for eventual sale to the public.

The rigidity of the grid, broken only by correction lines here and there to account for the curvature of the earth, contrasts with the irregular profile of the lake bed. The text of chapter 3, in which this map appears, emphasizes the centrality and dominance of Chicago in its region. Its running title, "The Metropolis of the Midwest," seems to be qualified somewhat by the grid of townships, because the lands of the city are reduced to simple townships, just like other locations in the region. One of the major shortcomings of the township and range system was its lack of centrality. Even the subdivision of the townships into thirty-six sections, each a mile square, worked against the development of local central places because such a democratic arrangement prohibited any section from being in the center.

The city of Chicago began as a creature of the township and range system. The first streets and blocks, laid out by James Thompson in 1830, were set in the southern part of section nine, rigidly following the surveyor's lines. As the city spread out to adjacent lands, the regular sectional division enabled the urban grid to be extended in orderly fashion across the prairie. Each section line became a major arterial street in the city, and other sectional divisions, like Division Street, would serve as major thoroughfares. If "order" was the watchword of the planners, no better principle could have been followed than to lay out the city and its region according to the format dictated by the sectional grid. Real estate developments started at some distance from the built-up area could later be smoothly connected to the city along the straight lines laid down by the initial survey.

Does an emphasis on the gridiron layout following the sectional lines, so much a part of this map, have any place in the overall schema presented by the *Plan of Chicago?* The answer is yes, on two counts. First, the structure of Chicago as it existed in 1906—which could be readily observed by the planners from their rooftop vantage point—followed the rectilinear pattern as far as the eye could see. It defined the existing city that Burnham and Bennett wanted to improve but not eliminate. Indeed, the rectangular arrangement of streets and blocks provided an efficient, understandable, infinitely expandable, and orderly structure for the urban fabric. The 1909 plan readily accepted the existing grid as the basic matrix out of which the new Chicago would emerge. All of the maps, plans, and views presented by Burnham and Bennett in 1909 should be viewed with this fundamental axiom in mind.[37]

What the rectangular grid failed to give the city—central places, radiating arteries, expansive views, and focal points—were precisely what the planners aimed to provide. The natural shoreline of Lake Michigan bestowed a sweeping curve to create a strong line of visual interest to the east. Beyond the city limits the "air line" routes of the various railroads and the natural slant of roads seeking the shortest route across the level prairies provided diagonal lines across the township grid. Remnants of early pioneer trails showed up as slanted streets even in the grid structure of the early Chicago. Deft planning simply needed to follow the web of the railroad pattern and connect or extend the radiating streets, expand the lakeshore from a thin line into a borderland, and create the missing central plazas and civic focal points. Then beacons of beauty would enhance the watchword of order. In a nutshell, the established grid provides one major key to understanding the *Plan of Chicago.* The arrangement of the illustrations in the book provides another.

The Plan as an Atlas

Viewing the *Plan of Chicago* as an atlas points to the cultural context of the proposal because one unlocks the secrets to books of maps by carefully noting their structure—how the individual maps are placed in a specific sequence—and by comparing the relative size or elaborateness of each image with the others. If this mode of analysis is used for the major plates in Burnham and Bennett's work, defining full-page illustrations as the major pictorial elements, a careful pattern seems to emerge.

The importance of the frontispiece cannot be overstated especially since the idea of starting with Lake Michigan seems to have come from Daniel Burnham himself. After viewing with great satisfaction the renderings from an aerial perspective provided by Jules Guerin, the master planner specifically commissioned him to use a similar approach to show the regional context of the plan with Lake Michigan as the major subject. The second illustration in the book provides a heading for page 1, the chapter on the origin of the plan. Although not a major plate by the full-page definition, it provides a temporal context to balance the spatial framework provided by the first illustration. Labeled simply as "Wood-cut of Chicago in 1834," it uses a vantage point off the northeastern edge of the lake port to provide a bird's-eye view of the small settlement (figure 6.3). The indistinct lines of the supposed woodcut look more like a charcoal sketch. But the message seems clear: Chicago began as a small town just seventy-five years ago, but the advantages of its location were already apparent in its active commercial life.

FIGURE 6.3. Woodcut of Chicago in 1834. (*Plan of Chicago,* plate 2)

Low frame buildings cluster about the mouth of the river to meet a series of sailing ships lined up along the newly constructed government pier. Several evergreen trees in the sandy foreground provide hints of the old Pine Street that would be transformed into Michigan Boulevard in the 1920s. The coming and going of the ships on Lake Michigan suggest that commerce was the chief reason for Chicago's existence. Warehouses and stores occupy Fort Dearborn's location; several of the ships are steamboats, bits of historical fabrication perhaps excused by artistic license. The lake, which takes up one third of the plate, furnishes a direct connection to the first illustration, while the settlement spreading along the shoreline suggests the possibility of continuing expansion across the prairie.

Immediately below the historical scene is another illustration, small in size, which embellishes the initial letter of the chapter. The device, suggesting a rich, presentation copy or a treasured medieval manuscript, is repeated in each succeeding chapter. Here the Court of Honor at the World's Columbian Exposition is sketched with the statue of Columbia in the foreground and the Peristyle in the background. On turning the leaf to page 3, the full page is devoted to a print of the same scene, at a greater distance, showing crowds of people taking in the spectacle. Another turn of the leaf presents a map of the world's fair grounds in 1893. Thus, by page 5, the careful reader must be impressed by the thoughtful selection and arrangement of the book's pictorial elements.

The second chapter, "City Planning in Ancient and Modern Times," begins with a photograph of the pyramids at Gizeh at the heading and a richly ornamented vase or loving cup as the initial letter decoration. The chapter is highlighted with twenty-five illustrations, but only one receives an entire page: a map in full color of Paris showing its transformation under Baron Haussmann. The facing page extends the idea with a second map of Paris, this time showing the proposals advanced by Eugène Hénard for additional radial avenues and circuit boulevards.

Chapter 3, "The Metropolis of the Middle West," starts with a map of the region cut into a circle with Chicago at the center. Radiating persimmon lines connect the hub with other cities

FIGURE 6.4. Chicago: General Diagram of Exterior Highways Encircling or Radiating from the City. (*Plan of Chicago*, plate 40)

in the region. The initial letter pictures a huge cottonwood tree near Chicago, standing alone on the prairie, dominating its surroundings. The next illustration in this section is the diagram of Lake Michigan set in the township and range grid discussed above. The chapter's only full-page print is a map of Chicago's immediate surroundings showing the highways radiating from it or arranged in concentric circles around it (figure 6.4).

Chapter 4, "The Chicago Park System," uses a miniature version of this last map to decorate the initial letter, suggesting a connection between the city's immediate surroundings and an appropriate park system. The illustration at the head of the chapter is a sketch made by Bennett of Grant Park from an aerial perspective. The view looks eastward from the West Side to the lake, showing the city under a blanket of snow and subtly revealing several major themes later advocated by the planners. It prepares a tantalizing first glimpse of what the *Plan of Chicago* is all about. Then, in spectacular fashion, the reader turns the page to reveal a magnificent double-page map (figure 6.5) tipped into the binding, displaying the proposed system of streets, boulevards, parkways, and parks in warm, inviting colors. This is the fundamental map in the book, and it is repeated in different scales, details, and colors to emphasize various aspects in several later plates (see especially plates 85, 103, and 110).

A series of smaller colored maps follow on the next three pages showing present and proposed parks in the metropolitan areas of Berlin, Vienna, and the District of Columbia. Each map reveals extensive green spaces, the obvious point of the series. But each one also shows a metropolis largely surrounded by countryside with its center in a landlocked position. In Chicago, by comparison, Lake Michigan cuts broadly through the urban fabric to place a vast expanse of water next to the very heart of the city. The lake presents Chicago with its greatest opportunity. "The shore of Lake Michigan," the text declares, "should be treated as park space to the greatest extent possible."[38]

The chapter continues with one of the greatest passages in the entire literature of city planning. In the sweep of Burnham's vision Lake Michigan itself becomes a great park, a natural resource so vital that access to its bounty becomes the natural right of every citizen:

> The Lake front by right belongs to the people. It affords their one great unobstructed view, stretching away to the horizon, where water and clouds seem to meet. . . . The Lake is living water, ever in motion, and ever changing in color and in the form of its waves. Across its surface comes the broad pathway of light made by the rising sun; it mirrors the ever-changing forms of the clouds, and it is illumined by the

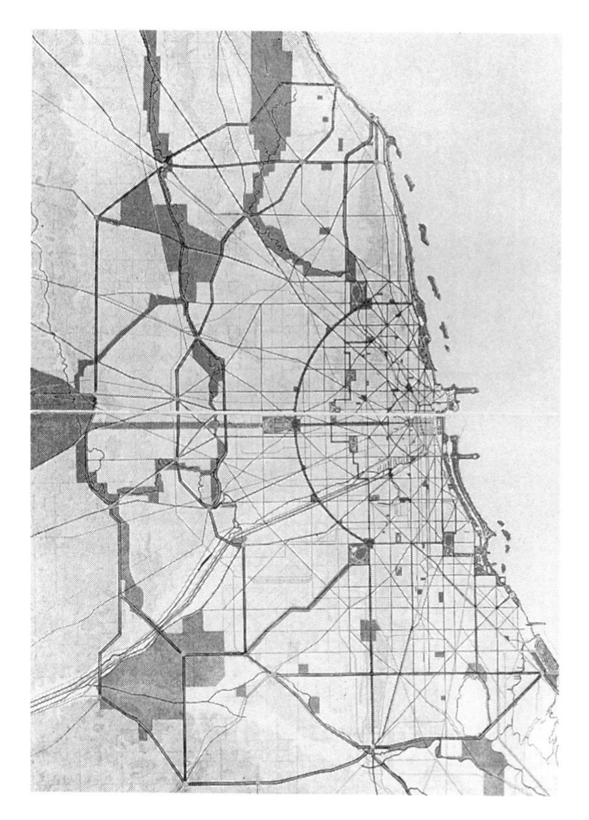

FIGURE 6.5. Chicago: General Map Showing Topography, Waterways, and Complete System of Streets, Boulevards, Parkways, and Parks. (*Plan of Chicago,* plate 44)

> glow of the evening sky. Its colors vary with the shadows that play upon it. In its every aspect it is a living thing, delighting man's eye and refreshing his spirit. Not a foot of its shores should be appropriated by individuals to the exclusion of the people.[39]

The paper opposite this passage is a blank sheet, seeming to highlight the point about "the great unobstructed view." Turning the leaf the reader encounters the first watercolor by Jules Guerin in the text proper. "View of the City from Jackson Park to Grant Park, Looking towards the West" presents the metropolis as a thin island stretching across the top of the page with the ivory paper continuing the color of the sky above and the lake below. The arrangement gives the impression of the broad expanse of the lake stretched across the lower two-thirds of the layout. The aerial perspective then squeezes the cityscape into a band of only an inch or two across the top of the print, reducing the topography of the city to a pleasing pattern of subtle colors. The only clearly distinguishable features are the parks, islands, piers, and harbor facilities proposed by the plan.

Immediately following this plate is another one which folds out to accommodate three detailed maps of the planned lakefront features. Part A reaches from Jackson Park on the south to Chicago Avenue on the north. Part B extends the treatment from Chicago Avenue north to Wilmette Harbor. The third image on plate 50 is a sketchy view of the South Shore park based on an early drawing by Daniel Burnham and Paul Lathrop. The following page presents a rendering by Jules Guerin of the same area looking in the opposite direction as well as a cross section through the park.

The maps on plate 50 provide a detailed guide to the impressionistic renderings of these proposed lakefront developments. The importance of these images to the presentation of the plan is suggested by the large size of the originals. The original Guerin watercolor for plate 49 measures over fifteen feet in length. The two maps that follow it together extended nearly twenty feet. They formed the major images when the plan was staged as an exhibit.

Chapter 5, a more technical discussion of transportation, uses a map of the proposed system for handling freight as a heading and a diagram of suggested inner city freight loops in the initial letter design. The three major illustrations in this section are maps, and so are most of the minor ones. Cartographic materials outnumber other types of illustrations sixteen to three. Two of the small features show traffic flows in scaled bars along various railroad routes, utilizing a map type used mainly by engineers. The three large plates in this section use printed maps as a base over which paint, colored ink, or pastels were applied. The first one, plate 73, is a topographic sheet issued by the United States Geological Survey. The other two use a base map of the plan's proposed arrangement of streets and waterfront facilities in the city center. A similar map, covering a slightly larger area and with more elaborate details, appears later as plate 110.

Chapter 6, "Streets within the City," begins with a discussion of the "natural features of Chicago," using a sketch of the central city as its head illustration. The bird's-eye view, looking west, gives half of the space to the lake, and is cut into an oval frame, suggesting a more natural shape as it points to Lake Michigan as the central geographic element in Chicago's location. The initial letter decoration shows two rows of stately American elm trees apparently lining both sides of a residential street.

The text in this section leads from a discussion of the city's topographic setting and its expansive views to a listing of the elements that make a city great. The first point is an "adequate means of circulation" and this naturally leads to a consideration of Chicago's streets. However, three double-page plates are tipped into the binding before the text gets to its discussion of streets, avenues, and boulevards. Once again, the nature and sequence of these illustrations suggests a purposeful arrangement.

The first two plates are detailed maps of the street pattern suggested by the planners. They elaborate the central portion of the initial presentation of the plan in chapter 4 (plate 44). Each map differs slightly in details but covers almost an identical area and serves a nearly identical function. The first is in black and white, but the second uses colors to highlight parks and arterial streets, with different shadings for present and proposed facilities. Thus the reader's artistic sensibilities are heightened as the pages are turned. But the reader would hardly be prepared for the dramatic view presented in plate 87.

Jules Guerin's bird's-eye view, "looking west over the city" from high above the lake, at sunset, uses a thin band of vermilion sky to define the horizon between city below and clouds above. Chicago looks almost monochromatic in chromium oxide green. The lake in the foreground, which receives almost as much space as the city, is given a lighter tint. This color value is continued in the Chicago River which clearly divides Chicago into three sections with its two branches. A similar shade is used for the Congress Street axis leading from Michigan Avenue to the proposed civic center plaza on the west side. Here the great dome of a new city hall marks the center of the urban ensemble. The very top of the dome is highlighted by a small patch of sunlight in Guerin's dramatic scene.

No mention of the civic center proposal would occur until the next chapter, and this small patch of light in the city at dusk is the first hint of the centerpiece proposed by the planners. From this tiny seed a great tree will grow, as the discussion of chapter 7 will soon reveal. However, before proceeding, another suggestive feature of Guerin's watercolor should be underscored: the way he extends the dramatic qualities of Chicago's lakefront into the Chicago River.

On all the maps and views up to this point, the Chicago River has been either entirely omitted, neglected, or treated in matter-of-fact fashion. It is almost impossible to find it on the previous large maps of the city and in the other illustrations as well. The only map where it is given a bit of attention, plate 75, features freight and traction circuits. The text follows suit. Only in a single paragraph on page 97 does the river receive some attention. "The Chicago River, which gave to the city its location and fostered its commerce," the text confesses "has become a dumping spot and a cesspool." Changes are imperative and "the opportunity should be seized to plan a comprehensive and adequate development of the river banks." Existing commercial uses would be accommodated "while at the same time the aesthetic side of the problem shall be worked out." After this statement little more is said in the text about the Chicago River, but it does appear more prominently on some of the following maps and views. As if to redeem this neglect of the river, Burnham and Bennett provided a single view of the river as a full-page, color illustration at the end of this chapter.

This view of the river looks toward Wolf Point where its two branches meet to form the main stem. The South Branch is the central feature of the print, but the title explains why it appears in the chapter devoted to "streets within the city." The view, it explains, shows "the suggested arrangement of streets and ways for teaming and reception of freight by boat, at different levels." The basic idea was to consider the river as an artery for freight and to write off any pedestrian access to it. The river banks could then be given over to wharfs and warehouse facilities with special lower level streets for teaming and the transfer of freight. Pedestrian ways and streets for vehicular traffic would be elevated above the river on a separate upper level.

This scheme would be developed more fully in the proposal for the Michigan Avenue bridge which would be either two separate bridges, one at each level, or a single, double-level bridge. A brief comment on extending this bi-level bridge idea to Wolf Point is provided in the caption to plate 106, a map which shows bridges crossing the three branches of the river and connected together to form a traffic circle. In contrast to Lake Michigan, the river banks did not belong to the people. Here one could not "escape from the petty things of life."

The idea for a traffic circle at Wolf Point almost surely came from *Études sur les Transformations de Paris* by Eugène Hénard, because a similar example is reprinted from this classic on city planning as plate 96 in the *Plan of Chicago.* The following four plates, also from Hénard's work, show "theoretical diagrams" for the streets of Paris, Moscow, Berlin, and London. Much of the text in this chapter is devoted to the value of diagonal and circuit streets in improving urban circulation patterns.

Plate 91, the final major illustration in chapter 6, uses a printed base map to show, in coded colors, the existing and the proposed arterial thoroughfares. The caption erroneously describes the colors, a mistake corrected on the errata sheet that accompanied some copies of the book. This map is interesting because it does not indicate the proposed lakefront developments, the sole example of this omission in the entire book. Perhaps these signs of haste indicate that this map was an afterthought inserted late in the process of production.

The Heart of Chicago

Chapter 7, "The Heart of Chicago," is also the heart of the plan. The central business district was where the project was conceived, developed, financed, and presented to the public. The chapters on parks, transportation, and streets all eventually lead to the center of the city. When viewed as an atlas, the *Plan of Chicago* places unusual importance on chapter 7. It is easily the most important part of the volume, receiving the most lavish illustrations and as many major plates as the rest of the chapters put together. The often repeated criticism that the *Plan of Chicago* focused on the downtown area, to the neglect of neighborhoods and other areas, is certainly supported by this analysis.

The method used by Burnham and Bennett to present their ideas for "The Heart of Chicago" was to repeat several visual themes again and again and to do so in a variety of different formats. The argument includes seven major renderings by Jules Guerin and all of the work contributed by Fernand Janin. Several other pieces specifically commissioned for the plan also appear in this section, including two that can be attributed to Chris U. Bagge.

The chapter heading is a sketch of the proposed plaza at Michigan Avenue and Congress Streets with the proposed Field Columbian Museum serving as the landmark structure at the eastern end of the wide expanse of pavement. The initial letter decoration is a sketch map of the plan's proposed restructuring of the center city. A reader would immediately recognize it as an image that had been presented in many previous illustrations, including four major plates (74, 80, 85, 86). The combination of map and view on the opening page repeats a pattern used in every chapter specifically presenting the plan's proposals.

At the turn of the leaf the authors provide a full-page detailed map (figure 6.6) of the image sketched in the initial letter. It has a serious tone, black ink on a light ochre tint. Only major streets and a few other features are labeled. The same map appears on the next page but this time the lake and the proposed harbor improvements are given a different wash and various colors are used on land to indicate parks, railroad facilities, and streets to be widened or cut through the city to implement the plan. No other picture in the book illustrates so well the extent to which existing buildings would be razed and new streets would be cut to fully develop the planners' vision.

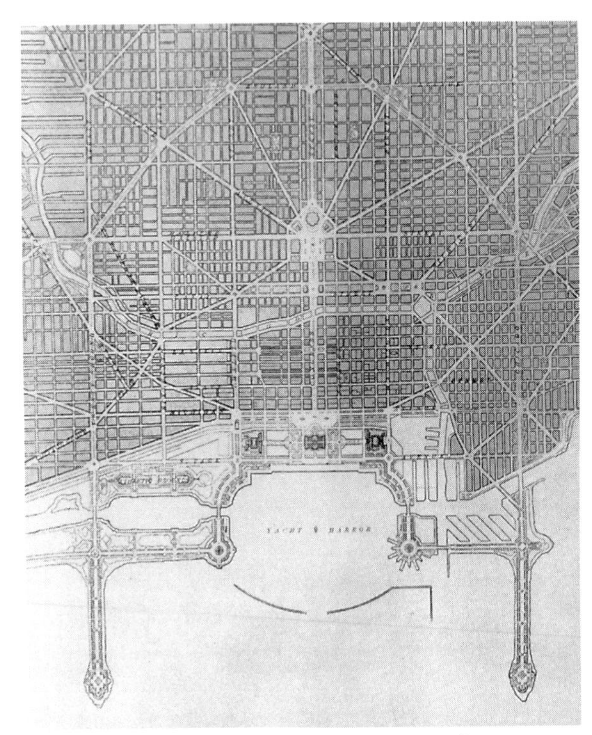

FIGURE 6.6. Chicago: Plan of the Complete System of Street Circulation. (*Plan of Chicago*, plate 110)

The next view, however, indicates the glorious boulevards that would result from this urban remodeling (figure 6.7). Jules Guerin's rendering of Michigan Avenue extending northward across the Chicago River starts with the existing Chicago Public Library building in the foreground.[40] This is the only case in all of the views of the plan where a contemporary building can be recognized. Perhaps it was meant to be a visual anchor, or to set a scale, or to provide a psychological tie between the present and the future city. The street scene presents a broad boulevard, spacious enough to accommodate plantings, a fountain in the center, and other decorative elements.

FIGURE 6.7. Chicago: Proposed Boulevard to Connect the North and South Sides of the River. Painting by Jules Guerin. (*Plan of Chicago,* plate 112)

Lake Michigan appears in the distant background, and the topography of the cityscape seems to repeat the level stretches of water and the prairie lands, "without hills or any marked elevation." The image recalls this comment made earlier in the text and seems to suggest how much inspiration Burnham drew from the natural setting of Chicago. The geography set a scale "seemingly without limit" where the city could expand endlessly. "Always there must be the feeling of those broad surfaces of water . . . [and] the sense of breadth and freedom which are the very spirit of the prairie." Here "no one should hesitate to commit himself to the largest and most comprehensive undertaking."[41]

The next major illustration, only a page away, uses a two-page spread to accommodate a strip map of the Michigan Avenue proposal. The image extends from Chicago Avenue on the left to Twelfth Street on the right. The map thus has east at the top. Immediately below the map, occupying some of the space left blank by the narrow strip map format, is a bird's-eye perspective of the same scene attributed to Chris U. Bagge.[42] The view, however, looks at a scene over the lake, thus reversing the directions of the map and making the intended connections between map and view very difficult. This is the one major shortcoming in the coordination of maps and views in the book. Another anomaly in the juxtaposition of these two images is the use of the existing Art Institute as a point of reference on the map in the place of the Public Library. The proposed widening of Michigan Avenue into a double street, one for local traffic and one for through traffic, would cut into Grant Park, leaving the Art Institute as an obstruction to traffic. On Bagge's view, however, the Art Institute disappears to make way for an unobstructed thoroughfare. This may explain why the building, so admirable an example of City Beautiful principles, is largely absent from the book.

Although several smaller illustrations are used to portray the development of Michigan Avenue north of the river, the major attention in the rest of the chapter is the expansion of the central business district south and west of the existing city center. The major extension of Chicago's downtown area northward along the new Michigan Boulevard was not anticipated by the planners who advocated instead the use of Twelfth Street (now Roosevelt Road) as a major axis of expansion. Thus, the intersection of Michigan Avenue and Twelfth Street is featured in three smaller sketches and anchors a double-page view by Jules Guerin (plate 121) picturing

FIGURE 6.8. Details from Chicago: Elevation of Grant Park and Harbor: The Eastern Facade of the City on Michigan Avenue and the Dome of the Administration Building of the Civic Center; and Chicago: Section Looking North, Taken through Proposed Grand Axis of the City. (*Plan of Chicago,* plates 125 and 126)

a magnificent cityscape to be developed along this axis from Michigan to Ashland Avenues, a distance of over two miles.

A similar painting by Guerin (plate 122) portrays the extension of the central business district west of the South Branch of the Chicago River, starting with a grand space for railway stations between Clinton and Canal Streets and terminating in the magnificent civic center at Congress and Halsted Streets. This is the first illustration to clearly present the focal point for the plan of Chicago that had been hinted at on plate 87 in chapter 6.

The following six major illustrations use the civic center and the proposed Field Columbian Museum at the east end of Congress Street to develop the theme of Congress Street as the major axis for the central city. Plate 124 from a drawing by Fernand Janin, is a two-page map of Grant Park and the Yacht Harbor with the museum as the anchor tying the central business district to the lakefront pleasure area. The grand building to feature the natural sciences at the foot of Congress Street is balanced by complementary buildings: a library at Twelfth Street and an arts complex at Washington Street. Passenger steamer landings and public promenades reach out into the lake toward a long breakwater which encloses an extensive yacht harbor.

The subsequent two plates (figure 6.8, details) require fold-out pages to present the facade of the city as seen from the breakwater (plate 125) and a cross section of the central business district along Congress Street from the lakefront to Halsted Street (plate 126). Both perspectives use the natural sciences museum and the civic center as visual markers to define the proposed cityscape. The ochre tones used in these sketches and the small scale of the renderings tend to surround the city with a fog, covering in suggestive tones the clarity one normally expects in cross-section diagrams.

The illusive quality of the proposal is repeated in plate 127, a painting by Guerin of the harbor scene presented on Janin's plan a few pages earlier. Here the painter uses a restricted and subtle pallet to catch the dream of the future city at dusk on a gray day. The next plate, also a two-page illustration by Guerin, shows the plaza extending from Michigan Avenue to the Museum of Natural History in Grant Park, filled with people on a sunny summer afternoon.

Plate 129 (figure 6.9) pulls together this sequence of illustrations, presenting a two-page map of the central city from the breakwater on the east to the civic center on the west, and from the main stem of the Chicago River on the north to the Twelfth Street corridor on the south. The detailed layout of streets and alleys gives the map authority, the gray ink on the ochre background provides a dignified look, and the extent of the proposed developments stimulates an exciting response. Michigan Avenue marks the center of the map at the fold line, with the entire right-hand page presenting Grant Park with its centers of culture and the yacht harbor beyond, thrust a mile out into the lake. The left-hand page divides the city into two segments with the South Branch of the Chicago River boldly marking the boundary. To the east, the traditional urban grid predominates; but west of the river the great plaza of the civic center creates a hub sending diagonal streets across the checkerboard in every

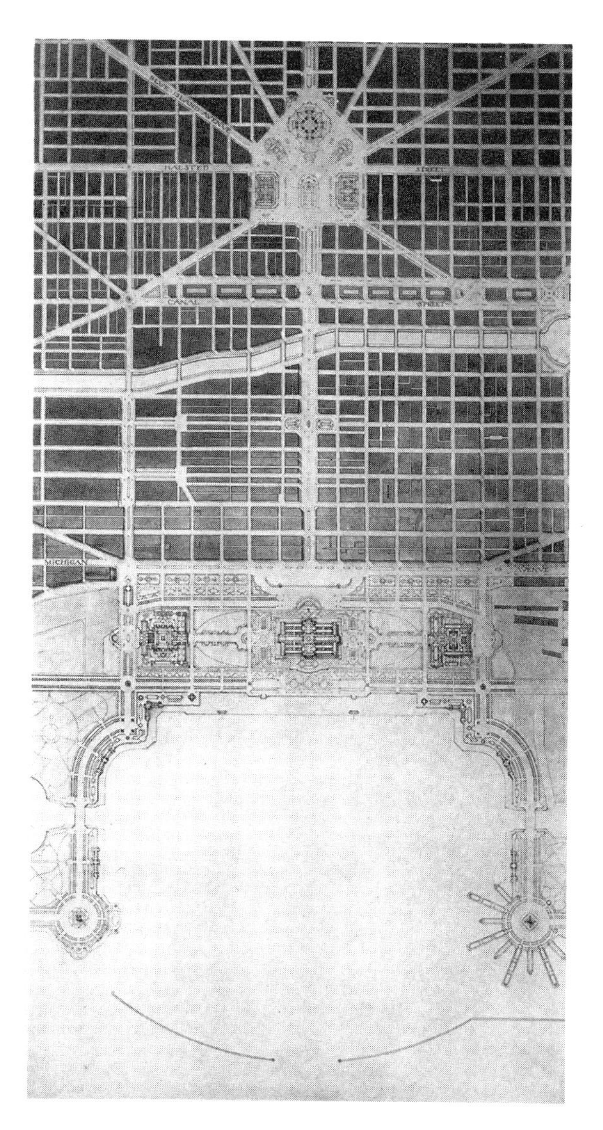

FIGURE 6.9. Chicago: The Business Center of the City. (*Plan of Chicago*, plate 129)

direction. The broad swath of Congress Street connecting the civic center with the central feature of Grant Park provides an east-west axis to balance the north-south course of the river.

The next three plates develop the proposal for a great urban focal point. Plate 130 is an enlarged

FIGURE 6.10. Detail from Chicago: Elevation Showing the Group of Buildings Constituting the Proposed Civic Center. Drawing by Fernand Janin. (*Plan of Chicago*, plate 131)

detail of the previous map, marking the extent of the civic center. It continues the suggestion of cartographic restraint, serving as a foil for the moving experience as the reader turns the page to catch a glimpse of Fernand Janin's drawing of the civic center extending over three pages in a fold-out plate (figure 6.10, detail). As the extra leaf is turned, the huge central structure is revealed in all its classical glory, taking up the entire central panel and suggesting what the planners meant when they commissioned this great dome "to mark the center of the city from afar" and to embody "the spirit of civic unity."[43] Plate 131 thus dramatically reveals the grandeur of the vision and establishes the focal point of the city. It also marks the climax of the visual display presented by the book. The illustrations after this carefully lead the observer down a path to civic commitment.

Jules Guerin's painting of the civic center just after sunset looks west to the broad expanse of the metropolis. It comes somewhat as an anticlimax, but this particular illustration, detached from the book, has become one of the most popular images of the plan of Chicago (figure 6.11). In the context of the presentation, however, Guerin's westward-looking view receives only one page in contrast to the two-page spread of its companion piece, looking east from above the civic center over the central business district to Grant Park and the yacht harbor (figure 6.12). Lake Michigan's placid waters fill the top half of the plate. Robert Bruegmann has carefully studied the development of this plate and concluded that it was originally designed for only one page but was enlarged because of its effectiveness as an individual piece.[44] Thus, the horizon line at the top was added to the original painting to give the plate the proper proportions for a two-page spread.

This keen observation helps answer two questions when considering the *Plan of Chicago* as an atlas or as a carefully arranged exhibit of illustrations. First, why should the view east receive two pages when the previous view to the west received only one? Second, why would the dramatic climax of Janin's drawing of the civic center be qualified by a secondary climax only a few pages later?

It is interesting to speculate that Burnham and Bennett, in this decision, probably erred on two counts. They not only qualified the rhetorical structure of their visual argument for a new city by enlarging the view to the east, but they also selected the wrong view to emphasize. For the view to the west, left in its single-page format, has proved to be the more popular image, arguably the most widely reproduced view from the book. Maybe this represents an American preference for a western orientation, toward the course of empire, or perhaps a favoring of the

FIGURE 6.11. Chicago: View, Looking West, of the Proposed Civic Center. Painting by Jules Guerin. (*Plan of Chicago*, plate 132)

glowing sunset rather than the cold, grey day of the painting to the east when the cityscape is dusted with snow. Or, pushing speculation to its limit, maybe succeeding generations of Chicagoans have preferred the view that looks out from the center city to the surrounding neighborhoods and suburbs which made up the vast expanse of the cityscape, basic building blocks of the metropolis which are largely overlooked in the plan.

Chapter 7 concludes with two additional sketches of the monumental classical building in the civic center. Plate 139 uses an entire page to frame an oval sketch of the plaza and surrounding buildings crowned by the central dome. The final illustration, placed at the end of the text, reproduces a study for the dome of the proposed civic center by Fernand Janin.

Chapter 8, a brief six-page concluding summary of the plan, shifts the emphasis from the illustrations to the text. It contains only three pictorial elements, all seemingly required by the overall design of the volume. The headpiece suggests a crude woodcut to match its companion in chapter 1. It begins in what might be a railroad yard or an industrial complex on the west side of the city, looking toward the landmark of the civic center dome with the bright sun rising behind it.

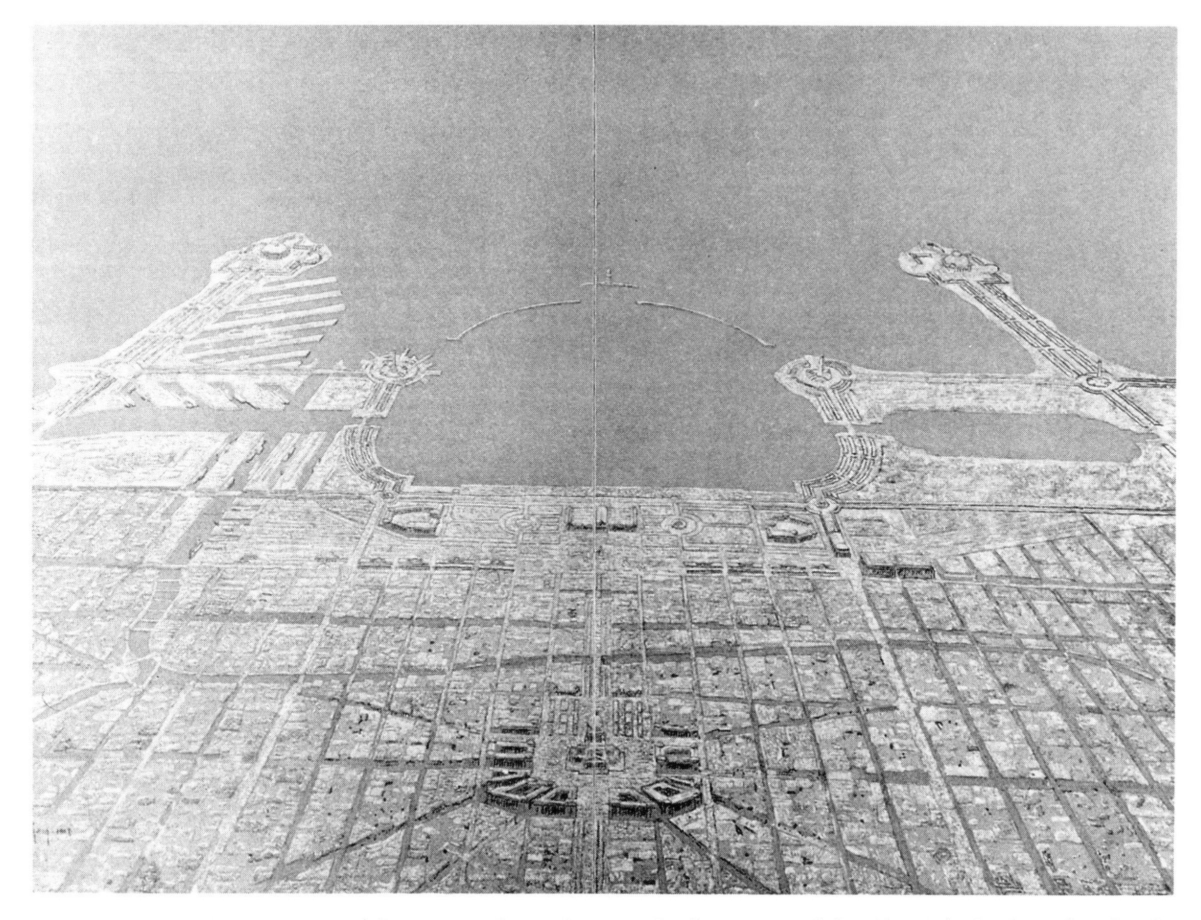

FIGURE 6.12. Chicago: View of the Proposed Development in the Center of the City. Painting by Jules Guerin. (*Plan of Chicago*, plate 137)

The initial letter reverses the direction, showing the eastern portico of the natural history museum with the great dome of the civic center looming in the distance behind it.

After this, a polished text, uninterrupted by any illustrations, reviews the principal elements of the plan, enumerates the cultural and economic foundations on which a great city might be erected, and makes the case for public commitment to the goals of the proposal. Two great gifts to the city are given as examples of the required civic pride and patriotic dedication. The Field Columbian Museum is pictured in the initial letter. The Benjamin F. Ferguson Monument Fund for erecting statues commemorating worthy individuals or important events in American history then furnishes the final illustration in the book at the end of the last chapter. But there may be more to the concluding plate than first meets the eye. Does it echo the rhetorical power of the frontispiece?

The text concludes with the axiom that the

FIGURE 6.13. The Great Lakes. From the Group by Lorado Taft. (*Plan of Chicago*, plate 142)

city should be a magnet. Yet Chicago's conditions in 1909 were "such as to repel outsiders and drive away those who are free to go."[45] The waters of Lake Michigan also eternally move away from Chicago to seek their outlet to the world ocean. True, they are constantly replenished by providential rains, but their ceaseless flow might also raise serious questions for the citizens of Chicago. This very issue was embodied in Lorado Taft's sketch of his proposed fountain "The Great Lakes" which concludes the *Plan of Chicago* (figure 6.13). Both the issue and the image also relate very directly to the frontispiece, Jules Guerin's bird's-eye view of the city on the shores of Lake Michigan. The argument of the planners thus begins and ends with the waters of the Great Lakes and their message for Chicago.

Taft's sculptured group of the Great Lakes, the first Ferguson monument, was not dedicated until 1913, four years after the publication of the *Plan of Chicago*.[46] Hence, it was necessary to use a preliminary study that differs slightly in details, but not in overall design, from the completed piece. The idea for casting the Great Lakes in a classical design reaches back to the Columbian Exposition. Burnham as the supervising architect expressed disappointment that the sculptures created for the fair did not portray the natural setting of the American landscape, especially the Great Lakes so near at hand, and on whose shores Burnham had spent almost his entire life. Lorado Taft, one of the young sculptors employed by the Exposition, may have remembered Burnham's words when he later used the Great Lakes as the subject for this large sculptural group to be set outside the Art Institute. Taft's address at the dedication ceremonies emphasized the extended hand of Lake Ontario at the bottom of the group. As the waters of the Great Lakes "escape from her basin and hasten into the unknown, she reaches wistfully after them as though questioning whether she has been neglectful of her charge."[47]

A careful reader of the *Plan*, searching behind every leaf to make all possible connections, will note that Taft's image follows the concluding paragraph of Burnham's text: "If, therefore, the plan is a good one, its adoption and realization will produce . . . [an environment in which] we and our children can enjoy and improve life as we cannot now do. Then our own people will become home keepers, and the stranger will seek our gates."[48] Unlike the inland waters that eventually migrate from the American interior, humanity will continue to seek the city on the shores of Lake Michigan. The brooding waters in Taft's art, and probably in Guerin's map as well (plate 1), were meant to suggest a conscience, a measure, and a sense of duty. The *Plan's* cartographic treasures, in the end, seem to gather the waters together, guiding them along familiar courses, evoking questions as well as appreciative thoughts, and ending with questions about the nature of the metropolis, the meaning of life, and the obligations of civic duty.

Notes

1. The oldest clearly identifiable map in the Western tradition is an image of a Neolithic town in Anatolia dating from about 6200 B.C.E. (James Mellart, *Catal Hüyük* [New York, 1967]). Jane Jacobs has used this site to reflect on the nature of cities generally in *The Economy of Cities* (New York, 1969).

2. On urban maps in general, see James Elliot's exhibition catalog *The City in Maps: Urban Mapping to 1900* (London, 1987).

3. Jürgen Schulz, "Jacopo de' Barbari's View of Venice: Map Making, City Views, and Moralized Geography before the Year 1500," *Art Bulletin* 60 (1978): 425–74.

4. Georg Braun and Franz Hogenberg, *Civitates Orbis Terrarum: The Towns of the World, 1572–1618,* with an introduction by R. A. Skelton (Cleveland, 1966).

5. I. N. Phelps Stokes and Daniel Haskell, *American Historical Prints: Early Views of American Cities* (New York, 1932).

6. John W. Reps, *Views and Viewmakers of Urban America* (Columbia, Mo., 1984), 172.

7. Once again John W. Reps is the major authority on the urban plans produced in the United States. See his *The Making of Urban America: A History of City Planning in the United States* (Princeton, N.J., 1965) and a handfull of later, more specialized, books. For discussions of various types of urban maps, see David Buisseret, ed., *From Sea Charts to Satellite Images: Interpreting North American History through Maps* (Chicago, 1990).

8. On the genesis of the Chicago plan, see Thomas S. Hines, *Burnham of Chicago: Architect and Planner* (New York, 1974), 312–20, and Daniel Bluestone, *Constructing Chicago* (New Haven, Conn., 1991), 182–204. A contemporary announcement of the plan is in the weekly published by the Chicago Commercial Association, *Bulletin* (November 2, 1906): 2:25, copy in the Bennett papers at the Art Institute of Chicago, box 7, file 2. For the rooftop workshop constructed for the planners, see Charles Moore, *Daniel Burnham: Architect, Planner of Cities,* 2 vols. (Boston, 1921), 2:14.

9. The best discussion of Burnham's career as a pioneer city planner in the United States is in Hines, *Burnham of Chicago.*

10. Hines notes that "hundreds of requests for information and professional advice left Burnham's office" to support the Chicago planning effort (ibid., 323).

11. The list of subscribers prior to June 1, 1909, is given on pages xvi–xviii of the *Plan of Chicago.*

12. Robert Bruegmann, "Burnham, Guerin, and the City as Image," in *The Plan of Chicago: 1909–1979,* ed. John Zukowsky (Chicago, 1979), 16–28, is the best discussion of the visual aspects of the book. Of the 142 numbered illustrations, Bruegmann notes that 81 were taken from other sources and 61 were prepared specifically for the project. Robert Bruegmann has kindly read this essay and has provided helpful comments, for which the author is very grateful.

13. Only three illustrations are directly referred to in the text. Two plates on railroad circuits and freight yards are cited on page 66; the street car routes and train stations on plate 80 are pointed out on page 74.

14. No study of the *Plan of Chicago* as a landmark of the printing arts has come to my attention. Kristen Schaffer's introduction to the Princeton Architectural Press's reprint of the *Plan of Chicago* (New York, 1993) starts by listing a variety of ways in which the book may be studied. Her own particular viewpoint utilizes the draft portions that were left out of the printed version to portray Burnham as a perceptive social critic.

15. On the nature of atlases, see Denis Wood, "Pleasure in the Idea: The Atlas as Narrative Form," *Cartographica* 24, no. 1 (Spring 1987), 28–29; J. B. Harley's comments at the Library of Congress Symposium in 1984 as reported in *Imago Mundi* 37 (1985): 89; and James B. Akerman, "Mercator's *Atlas* and the Origin of the Atlas Idea," a paper presented at the Newberry Library in January 1992. Wood quotes Mercator's first title page to make his point: "Atlas, or cosmographical medications upon the creation of the universe, and the universe as created." Akerman's definition of an atlas referred to "gathering, arranging, and designing maps from divergent sources into a coherent view of the world or some part of it in accordance with some intrinsic theme or concept."

16. Joan E. Draper, "Paris by the Lake: Sources of Burnham's Plan of Chicago," in *Chicago Architecture, 1872–1922: Birth of a Metropolis,* ed. John Zukowsky (Munich, 1987), 107–19.

17. Moore, *Burnham,* 2:147. See also Hines, *Burnham of Chicago,* 401, and Henry H. Saylor, "'Make No Little Plans': Daniel Burnham Thought It, But Did He Say It?" *Journal of the American Institute of Architects* 27 (March 1957): 95–99.

18. Daniel McCracken, "A Stitch in Time: Burnham's Chicago Plan Inspires No Little Tapestry," *Chicago Tribune*, January 24, 1988.

19. There are three good sources for biographical details. The most concise is Charles Moore's entry in the *Dictionary of American Biography* (1929), 3:302–7. His long, two-volume biography (see note 8) contains many relevant quotations from the sources. The 1974 study by Thomas S. Hines (note 8) is well researched and carefully organized.

20. Quoted in Paul Angle, "Chicago in 1856," *Chicago History* 4, no. 9 (Fall 1956): 272–73.

21. On Palmatary, see Reps, *Making of Urban America;* on his view, see Gerald A. Danzer, "Bird's-Eye Views of Towns and Cities" in Buisseret, *From Sea Charts to Satellite Images*, 145–55.

22. Carl W. Condit, *The Chicago School of Architecture: A History of Commercial and Public Building in the Chicago Area, 1875–1925* (Chicago, 1964), 43.

23. Burnham to his mother, May 11, 1868, quoted by Hines, *Burnham of Chicago*, 12.

24. For Wight, see Condit, *Chicago School.*

25. Root, like Burnham, has two biographers. The early laudatory work is by his sister-in-law: Harriet Monroe, *John Wellborn Root: A Study of His Life and Work* (Boston, 1896). The later, scholarly monograph, by Donald Hoffman, focuses on his work: *The Architecture of John Wellborn Root* (Baltimore, 1973).

26. On the Montauk and Burnham and Root's other buildings, see Condit, *Chicago School.*

27. A chronological list of buildings erected by Burnham and Root as well as the later D. H. Burnham and Company and its successor firms is found in Hines, *Burnham of Chicago*, appendix A.

28. Walter C. Behrendt, *Modern Building* (New York, 1937), 120, quoted in Condit, *Chicago School*, 69. In a way, the most stimulating recent scholarly analysis of Chicago architecture in the nineteenth century could be seen as a commentary on this text. See Daniel Bluestone, *Constructing Chicago* (New Haven, Conn., 1991).

29. Hines, *Burnham of Chicago*, 65.

30. Moore's *Burnham* reproduces a drawing by Willis Polk of a dozen buildings by the D. H. Burnham Company which groups them to form the skyline of an imaginary city (vol. 2, facing 206).

31. On the origins of modern urban design, see Mellier G. Scott, *American City Planning Since 1890* (Berkeley, Calif., 1969).

32. Guerin's life and career must be pieced together from a variety of sources, and not all of these agree. The best places to start are Bruegman, "Burnham, Guerin, and the City as Image," and John Zukowsky and Pauline Saliga, "Late Works by Burnham and Sullivan," *Museum Studies*, Art Institute of Chicago, 11, no. 1 (Fall 1984): 70–73.

33. Charles Moore, the industrious biographer of both Burnham and Charles F. McKim (who was in many ways Burnham's counterpart working in New York), has not attracted much scholarly attention. See his entry in the *Dictionary of American Biography.*

34. Joan E. Draper, *Edward H. Bennett: Architect and City Planner, 1874–1954* (Chicago, 1982).

35. Another version of Guerin's frontispiece was painted by Miles Sater for *Planning the Region of Chicago* by Daniel H. Burnham, Jr., and Robert Kingery (Chicago, 1956).

36. Not much has been written about Chicago's grid system of streets and blocks. Its clarity and simplicity seems to make it self-evident until one begins to look at it more carefully. A good place to start would be Mario Gandelsonas, *The Urban Text* (Cambridge, Mass., 1991).

37. For example, the two great piers which the *Plan of Chicago* proposed to flank each side of the center city are projected into Lake Michigan along section lines and terminate in circles which exactly mark the outer corners of sections 11 and 23 respectively.

38. *Plan of Chicago*, 50. See Daniel Bluestone, *Constructing Chicago*, to place the lakefront park idea in historical context, especially 185–190.

39. *Plan of Chicago*, 50.

40. The original watercolor for this plate is reproduced in color in John Zukowsky, ed., *Chicago Architecture, 1872–1922: Birth of a Metropolis* (Munich, 1987), pl. 77.

41. The full passage extends over four paragraphs on pages 79 and 80, *Plan of Chicago.*

42. Bagge, a member of the planning team, is given credit on page xiv of the *Plan of Chicago.*

43. Legend for pl. 130.

44. Bruegmann, "Burnham, Guerin, and the City as Image," 22–23.

45. *Plan of Chicago*, 124.

46. *Dedication of the Ferguson Fountain of the Great Lakes* (n.p., 1913). Timothy J. Garvey uses this speech extensively in his *Public Sculptor: Lorado Taft and the Beautification of Chicago* (Urbana, Ill. 1988).

47. Ibid.

48. *Plan of Chicago*, 124.

Contributors

David Buisseret is Garrett Professor of History at the University of Texas at Arlington. He has written extensively about the history of cartography, and edited *Monarchs, Ministers, and Maps: The Emergence of Cartography as a Tool of Government in Early Modern Europe* (1992) and *Rural Images: Estate Maps in the Old and New Worlds* (1996), both published by the University of Chicago Press.

Gerald A. Danzer, professor of history at the University of Illinois at Chicago, directs an M.A. program for teachers of history. He has written, edited, or contributed to a dozen books. The American Historical Association awarded him its James Harvey Robinson Prize in 1987 and 1990; his major area of interest is the use of maps as historical sources.

Richard L. Kagan is professor of history at the Johns Hopkins University. He has written extensively about early modern Spain, including particularly *Spanish Cities of the Golden Age: The Views of Anton van den Wyngaerde* (Berkeley, 1989), and is presently collaborating with the Spanish art historian, Fernando Marias, on *Urban Images of the Hispanic World 1500–1800.*

Naomi Miller is professor of art history at Boston University. She is the author of various books and articles on Renaissance Italy, including *Renaissance Bologna: A Study in Architectural Form and Content* (New York, 1989).

Martha Pollak is associate professor of art history at the University of Illinois at Chicago. She is the author of *Turin 1564–1680: Urban Design, Military Culture, and the Creation of the Absolutist Capital* (Chicago, 1991) and the editor of *The Education of the Architect: Historiography, Urbanism, and the Growth of Architectural Knowledge* (Cambridge, Mass., 1997). She is currently preparing a catalogue of Italian architectural treatises in the collection of the National Gallery of Art in Washington, D.C.

Nancy Shatzman Steinhardt is associate professor of East Asian Art at the University of Pennsylvania, and the author of four books on Chinese art and architecture, including *Chinese Imperial City Planning* (Honolulu, 1990).

Index

D

E

F

G

H